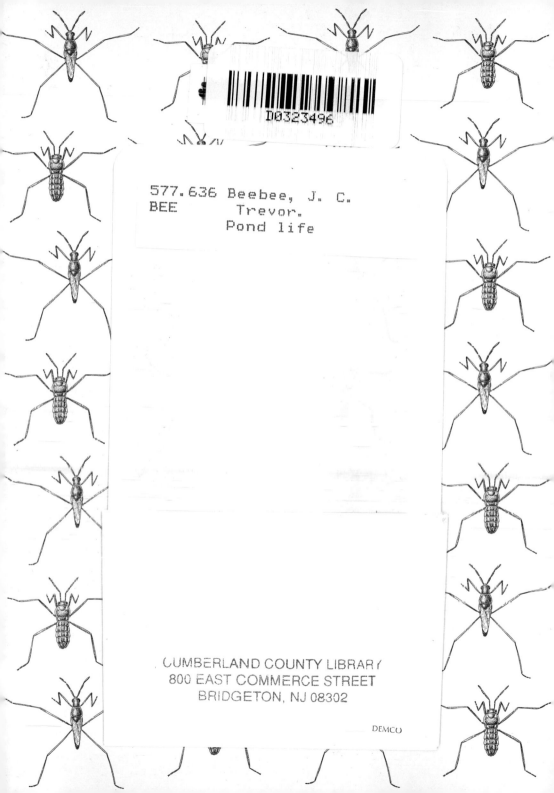

POND LIFE

POND LIFE

• TREVOR BEEBEE •

with illustrations by
PHIL EGERTON

Whittet Books

Endpaper illustration: pond skater and water cricket

First published 1992
Text © 1992 by Trevor Beebee
Illustrations © 1992 by Phil Egerton
Whittet Books Ltd, 18 Anley Road, London W14 0BY

Design by Richard Kelly

British Library Cataloguing in Publication Data

Beebee, Trevor J. C. (Trevor John Clark)
    Pond life. – (British natural history series)
    I. Title
    574.929

    ISBN 0 905483 99 5

Typeset by Litho Link Ltd, Welshpool, Powys, Wales
Printed and bound by Biddles of Guildford

Acknowledgments

Poking around in ponds is often rather a solitary pastime, but, like any other human activity, it benefits from contact with colleagues (some might say eccentrics) of similar bent. I am lucky enough to have a wife who enjoys a bit of aquaculture now and then, and children that tolerate it at least occasionally. I have benefitted from memorable discussions with John Clegg, and guidance from Garth 'water beetle' Foster; and more than ever from the enthusiasms of Brian Banks and Jonathan Denton, both of whom have forgotten the names of more pond animals and plants than I ever knew. My thanks to all these people for making a good hobby even better.

Contents

Preface

There must be something truly profound about the attraction of water to the human race. Certainly it goes far beyond the essential needs of drinking and washing. Unlike the horse, which may or may not drink when taken to it, try stopping a child racing in on the first day of a seaside summer holiday. Most of us will have started our fascination with water on the seashore, though the seaside holiday routine is surprisingly recent historically (it's actually been in vogue for only about a couple of hundred years). And, of course, most people don't live near the sea. However, few of us are far away from some form of freshwater body, be it an old canal, boating lake or the pond in the corner of the local farmer's field. Away from the seaside, bathing usually seems less attractive and the eternal fascination for water is more often fulfilled in other ways. One of these is trying to catch some of the many living things to be found in wet places. The wielding of a pond net, and the capture of newts, tadpoles and their like, is surely a paradigm of childhood in rural England. Even for adults, freshwaters often seem to hold as many attractions as the sea. Interest can appear as 'respectable' hobbies, such as sailing or fishing, but just as often large ponds and lakes are simply used as inland seasides to picnic near, or to sit around in deck chairs. Try to park near Frensham Ponds, in Surrey, on any summer weekend and you will see what I mean.

Water can inspire as well as fascinate us. Isaac Newton once remarked, late in his life, that he viewed his work as one who played on the seashore with all the unknowns of the universe crashing undiscovered in the oceans around him. In this book, the more modest ambition is to give an indication to the budding limnologist ('limnology' is the study of freshwater life) of what is there for the finding in ponds and ditches all over Britain. And this variety is enormous; there are more species of water beetles alone in Britain than there are resident birds. So this cannot be a catalogue of all the species you might find; its intention is to whet an already damp appetite, and serve as an introduction for those of a mind to sample one of the most accessible, and most rewarding, of wildlife habitats.

Trevor Beebee 1991

What is a pond?

This is not quite as obvious as it sounds, because wetlands come in a range of shapes and sizes and each has a characteristic wildlife associated with it. When, for example, does a pond get big enough to be called a lake? There is no universally agreed answer even to this simple question, but lakes are typically big enough to have a regular wave action, a deep area that never warms up even in summer, and lots of attention from windsurfers, water-skiers, anglers and the like. For our purposes, let's call a pond anything less than fifty metres (165 feet) or so across and preferably with no-one else around to notice your interest in it. This is arbitrary, and some places I would call ponds rather than lakes on the basis of their wildlife are certainly bigger, but it will cover most ponds in most places. It's also worth pointing out that size is no indicator of wildlife interest; some of the best ponds I have seen have been quite tiny, no more than 4 or 5 metres (12 to 16 feet) wide. Similarly, depth is not as important as you might imagine and shallow pools, perhaps because they get warm in the sun, can be among the best in the land. Indeed, temporary ponds or puddles that dry up completely in most summers (even English ones) harbour a characteristic wildlife including some of our rarest species (see Temporary ponds: a special case).

Another important factor is the type of soil on which the pond forms. Some, such as those in hard-rock mountain areas or on sandy heathlands,

Ponds don't have to be big to be interesting.

are poor in minerals and this is reflected in the water quality (which may be quite acid) and the rather limited numbers of plants and animals they sustain. In extreme cases there will be little life of any kind; to all intents and purposes these pools are just dilute sulphuric or organic acids, vinegar lakes by any other name. Ponds of this type are called 'oligotrophic' or 'dystrophic' depending on whether they are clear or stained brown with peat. Even in places as inhospitable as these, however, some creatures can survive. I know of one very acid pond where careful searching will reveal water spiders, dragonfly larvae and arguably the rarest water beetle in Britain. Most pools are of course on richer soils, such as clays or loams, and this will be obvious from the much greater variety of beasties these 'eutrophic' ponds contain. And at the other extreme from mineral-poor ponds are those subject to occasional inflows of seawater, which often stay much saltier than normal ponds but are still considerably 'fresher' than the sea. These are the brackish pools, which not surprisingly occur only near the coast or tidal stretches of rivers, and where, once again, we find a typical range of species not usually found elsewhere. Water from these ponds or ditches is often salty to the taste (if you are disposed to apply this simple but unappealing test), and on investigation may reveal marine shrimps as well as brackish specialists like the 'wasp-bellied' great diving beetle *Dytiscus circumflexus*.

A final word about ditches. To all intents and purposes a ditch is just a long thin pond, and I recommend you regard them as such. Their shape makes them easy to hunt in (there's no problem with getting to the middle) and to fall in (they're often quite steep-sided) and the plant and animal life in a good, neglected ditch ('dyke' in parts of England, 'rhyne' in Somerset) can be unbeatable.

Smells, stains and the mysteries of pond chemistry

A lot of odd things go on in ponds, and most of them are due to microbes, especially various weird and wonderful types of bacteria. Frantic decomposition of all the dead animals and plants accumulating on the pond floor releases some singularly noxious gases; methane is one, which occasionally bubbles up in quantity and can even catch fire at the surface to give a 'Christmas pudding' effect. Hydrogen sulphide is the 'bad egg' smell that results from digging your net too deep and fishing out a heap of sloppy mud; but in some ponds the generation of this gas by one lot of bugs is good news for another lot living above the mud, the so-called sulphur-oxidizing bacteria which convert hydrogen sulphide ultimately to sulphuric acid. In the normal course of events this sulphur is recycled into new life forms and no damage results, but one consequence of acid rain has been to overload ponds in susceptible habitats (so-called 'oligotrophic

ponds', with too few other chemicals to support much life) with sulphuric acid from the atmosphere; this is trapped by certain types of bacteria in the pond sediments, and released years later by others as high concentrations of sulphuric acid in a process that can probably go on for decades even after the acid rain stops (if it ever does) and keeps pH (a measure of acidity going from 1, very acid, though 7, neutral, to 14 which is very alkaline) as low as 3 in extreme cases. Fish or tadpoles die when pH drops to about 4, and food is digested in your stomach at about pH 2, which gives a feel for how inhospitable these ponds have become.

Completely natural processes can be pretty weird too; sphagnum moss, for example, is able to take in chemicals it likes (such as calcium) and in exchange chuck out hydrogen. This biological process makes the water around the plant very acid indeed, with pH dropping to 4 or below, a kind of in-house pollution that keeps potential competitors at bay. Another common feature of bog water is a distinctly brown colour; most of this is because of the very slow decomposition of plant material that goes on in poor, acid waters. It's not at all as bad as it looks, which is just as well because in some isolated Highland areas people still have to drink it. Perhaps the oddest-looking freshwaters are those, again usually on poor soils, that look like they are full of liquid rust. And rust it truly is, the work of yet another bunch of bugs, the iron-oxidizing bacteria. These cunning little beasts have learnt that energy is available not only from organic material that sustains most of the living world, but also (if you're equipped to use it) from some types of iron salts present in simple solution. The waste-product is an insoluble iron salt, ferric oxide (i.e. rust) and this is what they adorn their ponds and streams with. So this too is a natural feature, and not to be confused with real pollution that may be less unsightly or even invisible altogether but which causes genuine havoc among sensitive freshwater communities.

Canals: our longest ponds

The first canal ever built in Britain was the first I happened to meet as a boy, and if my experiences of them had ended there I wouldn't be including them in a book on pond life. The Bridgewater Canal was, and still is, bright orange for much of its length and those bits of it are about as appealing as an open sewer. However, like the railways, canals had their heyday in ages past and much of the country's extensive network has fallen into disuse (or, at least, into occasional leisure rather than commercial use). And a good thing that has been too, for wildlife and the pondhunter. If a ditch is just a long pond, a disused canal left to go its own way for a few decades is the long pond par excellence. Great places I have seen include parts of the Kennett & Avon canal in Berkshire, the Wey & Arun in Surrey and, especially, the Basingstoke Canal in Hampshire and Surrey.

This last one is particularly interesting, flowing slowly downhill from north-east Hampshire where it 'rises' on chalk hills, through the impoverished heathlands of western Surrey via a series of newly restored locks. All this means it has a huge range of water qualities, from hard to much softer, and thus a variety of wildlife unparalleled in any other water body I've ever seen. Not only is it outstanding for water plants and many water animals, but at one end is a long, disused tunnel that is now one of the most important bat roosts in southern England. Thirty years ago the Basingstoke canal had been disused so long that it was in danger of vanishing altogether; many of the old lock gates had rotted away, parts of the bank were breached and long sections were completely dry. The 1970s saw the start of a major effort to restore the canal, a mammoth task carried out largely by volunteer labour including skilled bricklayers to rebuild the locks. The result: a functional canal again, and a dilemma. One reason for restoration was to create a waterway for boat people and anglers, activities which can put at risk the wildlife if carried out too intensively. Let's hope a working compromise is reached such that this outstanding canal can continue to be all things for all people. That would be a heartening precedent many other canals could usefully follow in the years ahead.

Where to look for ponds

The slick answer is almost anywhere, and local children or a large-scale (1:25,000) Ordnance Survey map will quickly put you on the right track. However, some places are richer in ponds than others, and as we have seen above (What is a pond?) the type of pond you find may well depend on the kind of place you look in. So this matter does warrant a little serious thought.

The most obvious place to start, if only because it's within easy reach of many of us, is the nearest farmland. However, be prepared for disappointments with this tactic. Many farms are surprisingly 'dry' these days, with the spread of intensive arable farming (wheat doesn't need to drink from the field corner pond, but it can grow there and make a bit more money if the pool is filled in) and the installation of troughs to provide a more reliable and cleaner water supply for livestock. You can wander for miles in parts of East Anglia now with no opportunity to get your wellies wet, and even in the traditionally damper places such as the West Country and the Weald there are far fewer ponds than in the good old days. In particular, be prepared for many of those promising blue splodges on the Ordnance Survey maps either not to exist at all or to be just an overgrown damp hollow. Do persist though; there are still good farmland ponds, they just take more searching out than our predecessors had to put up with. Good advice is to select obviously low-lying areas, such as river valleys, which are often intersected with a maze of ditches and have a few proper ponds as well. The best of these, which occur in many counties all around the country, are among the most productive of all freshwater hunting grounds for the keen naturalist. I can recommend especially the ditches of east Norfolk, Kent, Sussex and Somerset for an inspired limnological experience on any warm spring or summer day.

Uncultivated land is also worthy of the pondhunter's attention. Heaths and commons often still have bogs and ponds, and although these may be nutritionally poor (see What is a pond?), they aren't always so, and if fed by springs or streams from richer soils can be very interesting indeed. Even the poor ones have their specialities, and merit at least a quick look. Sand dunes, too, can be excellent places for ponds. These, the so-called 'slacks', form in hollows between dunes and, although often temporary, can support unique plant and animal communities.

Mountains and moorlands are pretty wet places, and harbour good numbers of ponds, lakes and marshes. Often, though, these are uninspiring spots for pondhunters because they are not only low in nutrients but also rather cold. Nevertheless, the usual proviso applies: they have their own specialities, and the intrepid enthusiast will consider trekking across inhospitable terrain, often in trying weather, to catch a glimpse of them.

An interesting exception is the situation on chalk and limestone hills, which tend to be lower and in warmer parts of the country. Now, as any schoolchild knows, these are porous rocks and shouldn't support any ponds at all. The fact is, though, that fifty years ago some parts of the South Downs had as high a density of ponds as you could find anywhere in England. The reason: sheep farming. Sheep do well on downland turf, and to sustain them there the farmers of past centuries went to a great deal of trouble to create a vast network of so-called dewponds. These were holes dug out and lined with clay and straw or, more recently, with cement, to retain the water. In 1940 there was one per square kilometre in the Brighton area, but the decline of sheep farming sounded the death knell for most of them. Even so, a good few remain and, when found, are often very rewarding to the pondhunter because the surrounding soil is rich and the water in consequence supports a good variety of life. Too bad that finding the survivors usually means hiking for kilometres up staggering inclines past one dried-up relic after another.

History's villains: the drainers and builders

The British Isles of a thousand years ago were very much better off for wetlands than now, and not just because of the recent pond losses from neglect and so on. In those halcyon days, huge tracts of the country spent much of their time under water. We all learn how Hereward the Wake persevered as a thorn in the side of the Norman occupation by retreating, time and again, into the marshy islands and swamps of the Cambridgeshire Fens. Earlier still King Alfred had found similar sanctuary from the Danes in the impenetrable wetlands of the Somerset Levels. For centuries these and other low-lying parts of Britain were essentially no-go areas for Homo sapiens, but it was not to last. After many abortive efforts, things got serious in the seventeenth century when Dutch engineers (one Cornelius Vermuyden in particular) arrived on the scene and set about doing to England what they had previously done to Holland: making wet land

13

dry. The Cambridgeshire Fens, once covering much of Lincolnshire and northern Cambridgeshire, all but disappeared within a hundred years; similar was the fate of the Somerset Levels, converted into damp pasture intersected with elaborate systems of ditches and sluices that still, however, fail occasionally with dramatic consequences for those that live in these sub-sea-level localities. Rodmell and Southease, Sussex fishing villages in the Middle Ages, are now out of sight of the sea that once swept inland many kilometres beyond its present shore. All good news to farmers, no doubt, but the kind of history that brings a twinge of sadness to naturalists wondering what marvels these vast marshes once held. But it must be said that, except for the Cambridgeshire Fens, which have succumbed to intensive arable farming, what has replaced the old mires is not too bad. The ditches that cross these lowlands now are still among the most rewarding of places for pondhunting expeditions. Not so, unfortunately, for another mighty wetland that used to fringe the Thames where London now stands. As recently as the last century much of this remained intact and was home to quite rare species, such as great silver beetles, as we can see from the records and writings of naturalists at that time. Now only tiny vestiges remain, and most of these are under pressure for development of one sort or another. It will be a shame indeed if they also disappear, with their opportunities for enjoyment by an urban population otherwise so deprived of countryside experiences. Let's hope the planners of the late twentieth century will be a bit more tolerant of wet places than their all-too-successful forbears of the seventeeth.

Ponds and the law: keeping your nose clean

Knowing where a pond is may not mean that there are no more problems for the intrepid pondhunter. One thing we can be almost certain of is that the tempting, weed-covered waterhole will not be on land that belongs to its would-be investigator. Someone will own it, though, and by rights you should seek their permission to dip your net in it, thus avoiding accusations of trespass. There are, however, a few things which in practice amount to mitigating circumstances. If a public footpath runs near the pond (as they quite often do) there is nothing to stop you looking in, but even here putting a net into the water still requires permission on private land. So does climbing over a fence to get to the pond. On public commons there is usually little objection to pond dipping, but make sure there are no byelaws and especially that you are not on a nature reserve! I was once (quite rightly) harangued by an irate reserve warden for dipping in a pond which I thought was public, but in fact was on the edge of a County Naturalists Trust reserve. A frequent problem, not only on commons but also on farmland, is not knowing who the owners are and where to find them. Because casual investigation, done carefully, causes no significant damage to a pond there is little risk of legal eagles taking illicit dipping seriously, but there is always the chance of an uncomfortable few minutes at the sharp end of an aggrieved owner's tongue. My advice is really the obvious; find and ask the owner for permission if at all feasible before your net ever hits the water.

Care is also needed when you do start work, because some of the species you might meet are strictly protected under the Wildlife and Countryside Act, and because damage to or removal of any wild plants without specific permission is against the law. Critical in this situation is the intention of the offender; if what goes on is 'accidental', no prosecution is likely, but (for example) the deliberate catching of great crested newts will render you liable to prosecution. So arriving at a pond, netting at random and happening to catch a newt which is then immediately returned to the pond is OK; going to a pond you know to contain crested newts and using methods deliberately designed to catch them, without a licence, is not. Just how this difference in motive would be established in a court of law is of course another matter, and one which the responsible pondhunter should never end up in a position to test. The recipe is simple: take care at the pond to cause minimum disturbance to plants and animals alike, and return your catch to the water after you have finished inspecting it. Then slip furtively away.

What does a pondhunter need?

Without doubt the most important attribute is a thick skin, particularly if the pond is in a place where operations are likely to be observed by your fellow man. More seriously, there are a few items of essential equipment and of these a stout pair of wellies is perhaps the most obvious. I have dabbled in (and been drenched in) thigh-length waders, which can be useful in some situations but downright dangerous in others. It's all too easy to venture out too far and then get stuck in deep mud. I don't recommend waders for walking any distance to get to a pond, either. The most extreme case of cover-up I have come across was a pondhunter I once met kitted out in a complete wet suit, an eccentricity made all the more bizarre because his interest was in water snails – just about the slowest-moving of all pond animals and usually found near the bank anyway.

Next on the list has to be a decent, sturdy pond net. The flimsy items available from most pet shops aren't much good, and for a strong one you will have to find a specialist supplier or make your own. The best have a stout wooden or aluminium handle 1-2 metres (3-6 feet) long, an aluminium frame and a tough nylon-based mesh bag 20-30 cm wide and about the same depth. For most purposes I recommend a fairly coarse mesh size (at least 2mm); this will retain most animals but still move through the water without too much resistance. Dealers in swimming-pool equipment often have a good range of materials for pond nets, though of course they aren't manufactured for that purpose.

You now have enough equipment to get on with it, but there are a few useful ancillaries I should mention. A powerful torch, for looking in ponds after dark, is a valuable alternative to netting which can be very informative and much less energetic. Collecting boxes are useful if you want to bring specimens home, or even just for careful inspection on site; Tupperware-type plastic ones are handy. And a magnifying glass ($\times$ 10 or $\times$ 20) can assist with the finer points of identification when difficulties arise.

Finally, you will naturally want to know what the stroppy little beast with legs everywhere and enormous fangs latched into your finger actually is. While later sections of this book give some guidance on this, there are so many varieties of pond life that for precise identification you will often need more specialist tomes. It all depends how seriously you decide to take your pondhunting; if, like me, you will usually be content to know approximately what it is (for instance, whether it's the larva of a dragonfly or of a beetle) this book should suffice; if, on the other hand, you want to establish exactly which dragonfly larva is in the net you will need a specialist work showing how to distinguish the forty or so species. A guide to such books is given on p.122.

All kitted out and ready to go.

Pondhunters past and present

Searching out the wildlife of wet places has been going on for a long time in Britain, very much in line with the tradition of natural history pursuits that arose in the late eighteenth century. Not too much is known about the habits of the earliest pondhunters, though by the end of the nineteenth century the art was sufficiently well developed for authoritative books to appear (a couple of the best I know are cited in the bibliography). Since 1900 'limnology' has made enormous strides, and the path is beset with heroes of the age. Frank Balfour-Browne, for example, was a water-beetler or unparalleled single-mindedness. Much of his life seems to have been spent heaving beetles out of ponds and ditches throughout the length and breadth of Britain, culminating in three momentous volumes in which you can learn every detail of beetle anatomy known to man. It's easy to get the impression he did little else but eat, sleep and hunt water beetles. More recently, John Clegg has worked wonders in popular-izing the study of pond life with eminently readable books including the ones that aroused my own interest many years ago. Some things never change in pondhunting circles; whereas children think nothing about diving in with pond nets, adults occasionally become self-conscious about it and could benefit from one remedy advocated by John Clegg: disguise your pond net as a walking stick, with the necessary devices to make it collapsible and so stride boldly across the countryside with furtive intent undetectable by casual passers-by. Sadly, as far as I know no net manufacturers have developed this idea, so the only alternative is to take a few children along (the net is for them, you understand) if the occasional odd stare bothers you.

Another pioneer, this time of a more strictly scientific bent, was T.T. Macan. For many years Macan was a prime mover in the Freshwater Biological Association laboratories at Windermere, and with his colleagues led the way in serious research on how freshwater ecosystems function. To these and all the others who have boldly gone before, today's pondhunters owe a substantial debt. The general biology of all the common pond creatures has been worked out for us, and we also know many of the best places to go to find them. But there is still plenty to discover, and I don't doubt that in a hundred years' time pondhunter historians will be reflecting on more recent generations of dabblers in the art. I could name a few candidates, but discretion is the order of the day.

When to go

It's true to say that, except when a pond is covered by ice and thus inaccessible even to the most fanatical hunter, there is something to look for at almost any time of year. Of course many animals will be hibernating, or at least pretty inactive, during the winter months. Many pond insects, for example, bury themselves in the bottom mud during really cold weather and most vegetation dies back or disappears altogether. Even so, a mild spell in January or February can be very rewarding. Some beasts are tempted into activity, but are still sluggish; this and the lack of weed in winter makes catching them particularly easy. I have regularly found warm winter days the best for chasing marsh frogs, which in spring and summer become so agile it's difficult to get within ten metres of them. Newts, too, often arrive at ponds in mid-winter, and in the West Country common frogs will be spawning in January or February. A special word about temporary ponds: for them, winter is usually the best time of all, because they fill with autumn rains and dry out in spring or early summer. Those creatures adapted to such a precarious lifestyle, such as the fairy shrimps, are often at their most abundant in the darkest days of the year. So don't write off winter, just wrap up warmer and choose the mildest days for your excursions.

Spring is, without any doubt, the most exciting time of all for searching out pond life. The lengthening days trigger growth of water plants and, in almost all pond animals, the urge to breed. Pools in April and early May are still sufficiently weed-free as to make netting easy, but are veritably buzzing with activity. Adult beetles and bugs are at their most obvious; the newts are in, and many ponds will be teeming with tadpoles. Fish are tempted into the shallows for spawning and it's a time for surprises when, in a pond where you didn't even think there were any fish, an enormous female pike swirls ominously out from the water's edge. Water birds will of course be nesting, and care is needed to avoid unnecessary disturbance in sensitive areas. So if your time is a precious commodity, and you have to choose just one season for your pondhunting activities, make this it.

The dog days of high summer offer different opportunities for the keen limnologist. Weed will be thicker, often impossibly so for netting purposes, but the flowers will be at their most luxuriant and by now there will be more types of weed to look for. The amphibians will mostly have gone, and the fish retired to deeper and cooler waters. However, it's the best time to look for many types of insect larvae, which hatch in spring and grow rapidly to their maximum size by July or August. In this realm come some of the dragonfly nymphs, and the huge larvae of the biggest water beetles (see pp.61 and 70). Grass snakes will be around in the early summer, probably trying to catch baby frogs, and other reptiles as well as many mammals resort to water when the going gets hot. But perhaps best of all are the various flies (non-biting varieties) which emerge in the summer months from their aquatic larvae, and some of which then haunt the pond until the killing frosts of autumn. Mayflies have a brief, explosive showing, but prime viewing must be the glorious variety of damsel and dragonflies that patrol over the water all summer long. These are there just for the watching, no need for that net and the frustrating toil of the now dense weed beds. Indeed, pondhunting in summer can amount to little more than a well chosen picnic spot. The pièce de resistance at this time of year, if you are lucky and patient enough, is to watch a dragonfly emerge from its old nymph skin halfway up a reed or rush.

With the coming of autumn, the pace of life in and around the pond gradually slows down and ultimately the water develops its winter gait. There are one or two high spots, though. Those beetle larvae that pupated in mud through the late summer now emerge as new adult beetles and sometimes, in a particularly successful year, your net will come up with lots of the same species all at one go in an autumn outing. Predators know about these 'mass emergences' too, and it's not uncommon to find piles of beetle remains around ponds in September after herons, mink or the like have had a binge. Freshly filled temporary ponds are also rewarding, since their inhabitants appear and develop with quite astonishing speed.

The seasons, then, are the prime consideration with respect to planning your forays. Weather also matters, and the good news is that the nicer it is, the better your chances of a productive day out. Sunshine isn't just a human delight, but the warmth it brings stimulates activity (and thus increases the chances of being seen or caught) in most pond creatures too. Of course most people go ponding during the daylight hours, but, as mentioned earlier, searching with a powerful torch after dark can be fun. Much less enjoyable is plunging waist-deep into unchartered potholes in pitch darkness, so make sure your night visits are to places already familiar in daylight. Choose a warm, still night without wind or rain to disturb the water surface and go shortly after dark, before the temperature drops too far. Finally, make extra allowances for the panicking of passers-by confronted by a torch looming out of the darkness around the edge of a pond or marsh.

Pondhunting: the nitty gritty

Having decided where to go, when, and what with, it's a good idea to know what to do when you get there. What not to do is to rush straight in, wielding your net vigorously and scattering plants and animals in every direction. Far better to approach the water quietly, even stealthily, as if about to cast a fly for a wary trout. Make the most of unobtrusive observation; there may be mammals or birds, swimming or at the pond edge, that will certainly make themselves scarce if they see you first. Some of my best views of deer have been in circumstances like this, not to mention herons and kingfishers, early on a summer morning. Grass snakes, mink, water voles and shrews, even otters can be watched rather than precipitated into flight by the taking of due care, and in the frog and toad breeding seasons the congregations of adults can be observed going through all their croaking, fighting, mating and egg-laying rituals quite oblivious of a human presence.

Phase two should also be a rather passive affair, observation again but this time at rather closer quarters. Look carefully at the vegetation, and see what species of water plants are present; also look for obvious eggs, such as frog or toad spawn, newt or fish eggs attached to leaves in the pond, and possible bird nesting sites to avoid later disturbance. Just sit and stare at the water for a bit; you may well see fish, newts, and even the larger insects such as diving beetles coming up for air. Once the net goes in all this tranquillity will disappear, so make the most of it.

Finally, when you're satisfied there's nothing else to see just by looking, it is time at last to start using some energy. Use your net carefully, try not to damage the weed beds beyond what is absolutely necessary, and above all return what you pull out back to the water after you have finished. However carefully done, netting is sure to traumatize the pond and its inhabitants for a while. Tip your net contents onto as flat a surface as possible, and return gill-breathing animals (such as sticklebacks) to the water as quickly as you can. With a little experience you will soon get to know the main characters, and thus which to deal with speedily. You should also learn (preferably by reading rather than experience) which can give you a memorable bite, and how to handle them with the appropriate reverence. Anything of special interest can be housed temporarily in a

plastic box, with water of course, and returned after inspection and identification. Don't leave them in strong sunshine, since many species will die if they overheat, and try not to make the mistake of putting predator and prey together! Tadpoles boxed with newts or diving beetle larvae will become history within a few minutes, an interesting lesson on how quickly the predators' hunting instincts overcome the fright of being yanked out of their pond in the midst of a frantic netting session. Always return your catch to the same pond, and preferably the same bit of it; the animals and plants were adapted to live in that particular environment and may not survive being dumped somewhere else, however suitable it looks to the human eye.

Torching at night is rather simpler than all this, since it's more or less impossible to approach a pond flashing a light beam about without disturbing any peripheral wildlife. The alternative, sneaking up before you turn the light on, courts the disaster of the pond finding you first as well as the risk of accidentally treading on bankside inhabitants such as toads. Many of the truly aquatic animals, however, emerge from weed cover and venture into shallow water during the hours of darkness and are easily surprised by a sudden torch beam. Unfortunately so very often are nearby humans, and if there are likely to be any, you should be tactful about after-hours pondhunting. Torching is particularly good for great diving beetles, great crested newts and some types of fish but has the disadvantage that identification must be in situ. Trying to wield a torch and pond net at the same time is the best way I know of getting very wet indeed.

Twitchers at the pond

Tearing around the country in response to the sighting of a rare species or unusual visitor has become endemic among birdwatchers in recent years, largely because communications are now so staggeringly efficient in the ornithological world. One glimpse of a lesser spotted humdinger in Scotland before breakfast is old news in England by lunchtime, with traffic jams on the M1 as enthusiasts flock to the happy scene. Pondhunters can't quite match that, but a goodly number are prepared to travel to find beasts of special interest. It took me many years and a long way from home to find my first great silver beetle, an event still emblazoned clearly on my memory. The hunt for rare species has a curious fascination for many naturalists, I suppose because of that extra satisfaction when the quarry is finally tracked to earth (or pond in our case). It also involves visiting some particularly attractive parts of the country, from the Somerset Levels right through to the Scottish Highlands; all have

their special rarities that provide an excuse for trekking there. My own hit list of rarities worth a trip to see includes the lesser silver beetle (Hydrochara caraboides), the medicinal leech, several of the rarer great diving beetles and the tadpole shrimp. The latter is known regularly only from a single pond in the New Forest, though it was found as far north as Scotland earlier this century; it is a particularly tantalizing inhabitant of temporary ponds that could crop up almost anywhere, but at the time of writing has still eluded me. Other pond twitchers might be keener on rare plants, or dragonflies; there's plenty of scope, with so many different types of life making their homes in freshwater. It should of course go without saying that pursuit of rarities must be matched with a sense of responsibility about their safety and conservation; success should normally be followed by careful handling and release, with minimal disturbance to the species in question and its habitat.

Pond life at last: the plant communities

First impressions of a pond – like any other habitat – come from looking at the plant species in and around the place. First impressions are important, too, because with practice a scan of the vegetation will give you valuable clues about the likely richness of the pond in terms of animals you might find there. In other words, it's a rapid guide to whether you should hang around and look more closely, or go and find somewhere else that is more to your taste.

The first things to notice are the pond surroundings. Is it largely shaded by trees, or mostly open to the sun? Ponds with heavy shade will naturally be cooler, and less light means less growth of aquatic plants. This in turn means less productivity, and in most cases less overall interest. The classic 'shady' pond is shallow, dark, and with a smelly (even by pond standards) bottom mud rich in dead leaves. Often this condition represents the end of a pond's life; after all, many trees such as willows and alders like to grow in damp places and over the years will overwhelm any pond unless there is management, or grazing animals, to stop them. Even if unshaded by trees, there may be extensive growths of reeds around the pond edges, or along a ditch, that make the water inaccessible and uninviting. This situation too represents a late stage in the life of a pool, as encroachment by reed beds will eventually turn the pond into a swamp and ultimately dry land. Finally, if you are on a heathland or similar common, or an upland moor, you may find ponds with lots of sphagnum mosses around and on their banks. This is a sure sign of acidity; unless you are after an acid specialist you would do well to move on, since acid pools are notoriously poor in terms of wildlife diversity. Not many things like to live in vinegar.

What is on the pond is also an important guide to character. Many types of floating plants exist (see The floaters, p.34) but keep your eye on one in particular: the lesser duckweed, *Lemna minor*, is another bad sign when present in dense, bright green mats over the whole pond surface. It usually indicates a species-poor pond, either with stinking mud and large amounts of rotting matter beneath the water, or pollution by run-off from nearby farmland that contains too much artificial fertilizer. Some tree shade and this duckweed quite often go together, and I usually give ponds like this little more than a cursory glance. However, this isn't always fair,

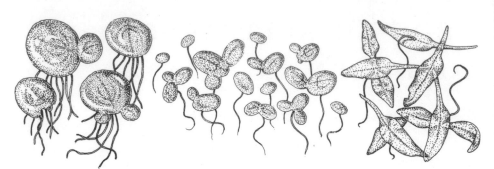

Greater (left), *lesser* (middle) *and ivy-leaved* (right) *duckweeds. All are about three times life size.*

since some lesser duckweed – especially if it is in a mixture with other duckweeds – can indicate an interesting site, so don't be too hasty to dismiss a green carpet. In my pondhunting apprentice days I ignored a pond of this kind for a long time, to find later that it was the only place in the area with water spiders. Another undesirable (even disgusting) spectacle is a pond covered with thick, slimy blanket weed; this green alga, when present in such excess, usually indicates pollution by silage, cow dung or something similar and merits the widest possible berth.

Undoubtedly the most encouraging sign is the presence in a pond of one or more species of plants growing mainly or completely underwater, the so-called oxygenators or true aquatics, or some of the other floating species such as frogbit and ivy-leaved duckweed (see following sections). There is a considerable variety of such plants, each no doubt with its particular growth requirements, but frequently several types will grow in the same pond or ditch together. These plant communities are typical of rich, unshaded and unpolluted freshwaters and are the clearest signal to stop right there and start looking seriously for what lurks beneath.

Taxonomy: sorting out the relatives

Classifying animals and plants and working out who is related to whom is one of biology's longest-standing activities. It's also had a poor image in some quarters lately, with the notion abroad that it's an old-fashioned and lonely pursuit of backroom boys in dusty museum archives. In fact taxonomy remains as important as ever, and with modern methods ('DNA finger-printing' and the like) is far from

'FAMILY TREE' OF COMMON TOAD, *BUFO BUFO*

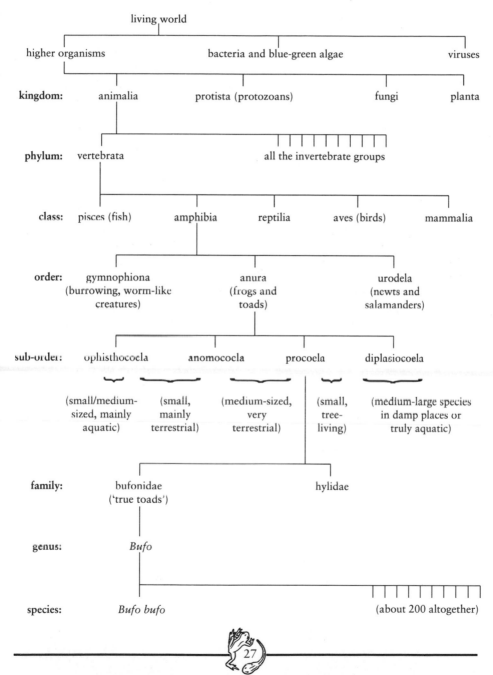

living world

| higher organisms | bacteria and blue-green algae | viruses |

kingdom: animalia — protista (protozoans) — fungi — planta

phylum: vertebrata — all the invertebrate groups

class: pisces (fish) — amphibia — reptilia — aves (birds) — mammalia

order:
gymnophiona
(burrowing, worm-like
creatures)

anura
(frogs and
toads)

urodela
(newts and
salamanders)

sub-order: ophisthocoela — anomocoela — procoela — diplasiocoela

(small/medium-
sized, mainly
aquatic)

(small,
mainly
terrestrial)

(medium-sized,
very
terrestrial)

(small,
tree-
living)

(medium-large species
in damp places or
truly aquatic)

family:
bufonidae
('true toads')

hylidae

genus: *Bufo*

species: *Bufo bufo*   (about 200 altogether)

27

ready for consignment to the scientific history books. All scientific classification is based on the brilliant idea of Linnaeus, over 200 years ago, to introduce a simple binomial system using a double Latin name for each and every species known to science. On top of this comes a kind of pecking-order in which species are grouped together in ever-vaguer combinations. Illustration by example is crucial here: the English common frog has the binomial Latin name of Rana temporaria, which means literally 'temporary' or summer frog. Of course the common frog in England is not the same species as the common frog of (let's say) the United States, so here is another good reason for using Latin names: they avoid confusion as to what's really what. Even so, North American frogs are very similar to ours and this is taken care of by putting them in the same 'genus'; for example, we find in the United States such beasts as the wood frog (Rana sylvatica) and the leopard frog (Rana pipiens), sharing the same genus name (Rana) but with different second, species names. So biologists were perverse and did the opposite of human families, which keep the last name the same and vary the first.

But then we can also say that all these frogs are tailless amphibians, and so are toads; this level of similarity, obviously a bit less close than the genus, is called an Order. All frogs and toads in the world are therefore in the same Order, known as the Anura. Then again, frogs and toads have a lot in common with newts and salamanders; all have moist skins without fur, feathers or scales and most lay their eggs in water to produce tadpoles. So all these animals belong to the next grouping of the taxonomic hierarchy, this time a Class: the Amphibia. And we haven't finished yet: amphibians have distinct, segmented backbones just as do fish, reptiles, birds and mammals. So all five of these groups are classified as the Phylum of Vertebrates, showing among other things that if you look hard enough frogs and people must be distant relatives. Finally, the vertebrates take their place in the kingdom of Animalia, together with insects, snails, and everything else that can reasonably be called animal. The living world has five kingdoms at this apex of the taxonomic tree; the others are plants, fungi, protozoa and bacteria. Sooner or later everything is related to everything else, it's just a question of whether it's borther and sister, or cousin two thousand times removed. Taxonomy aims to put the whol living world into perspective, all based on the idea that everything evolved from a single, unimaginable ancestor some three billion years ago.

Plants of the pond edge

Since all plants need water, it's not too surprising that many species like to grow close to a reliable supply of the stuff. This makes for all sorts of difficulties in defining what is really a pond plant, what is a bog or marsh plant, and what just likes a bit more dampness than the average petunia. The most well known denizens of the pond edge are undoubtedly the reeds and rushes, of which the enormous bulrush (*Scirpus lacustris*) and the false bulrush or reedmace (*Typha latifolia*) are formidable examples. The latter, with its distinctive sausage-shaped brown seed pods, is thought of as the bulrush by most people almost entirely as a result of a famous painting by Alma-Tadema of Moses in a basket surrounded by the false variety. Standing higher than a man, vast beds of bulrushes can be found around some of the larger ponds and lakes of Britain. Smaller, and usually even more abundant, are related plants such as the common reed (*Phragmites australis*), the burr-reeds (*Sparganium* species) with their conker-like fruiting bodies, and the sedges (*Carex* species) which look rather like proper grasses with solid stems. Sedges often grow around as well as in the water, and, like reeds, can form large impenetrable beds.

The true rushes ('*Juncus*' species) have a spiky, hedgehog look and are widespread not just at the edges of ponds but also in damp meadows and

False bulrush (left) *and sedges* (right).

29

Rushes (Juncus) (left) *and cotton grass* (right) *in a typical wet moorland scene.*

other wet places. Reeds, sedges and rushes run to many species and specialist keys are needed to identify one from another. Reed beds form both a unique habitat in their own right and also the basis of one of the few industries associated with pond life. The cutting of reeds and sedges has long supplied local thatching operations, and provided vast quantities of winter bedding for domestic animals, especially in the fenlands of East Anglia. For the pondhunter, their main interest is as a safe haven for birds such as bitterns and marsh harriers, though they can be good places for otters and in Norfolk offer the best chance to view the rare swallowtail butterfly. They are also ideal places in which to get completely lost, being both featureless and with vegetation so tall that spotting landmarks is out of the question. Be warned.

Some other notable plants of the grass and sedge families are also often associated with water: float grass (*Glyceria fluitans*), with its leaves often forming a dense covering on the water around the pond edges, is typical of rich pools; while purple moor grass (*Molinia caerulea*) and cotton grass (*Eriophorum angustifolium*), with its tufty seed heads, adorn the more acidic ponds and bogs of heath and moorland. *Molinia* in particular can be rampant, forming large areas of tussocky growth which, by the way, scarcely ever look purple. This is a plant which seems to have benefitted from pollution, and there is concern about its spread on heathlands, enriched by nitrates from acid rain, at the expense of heather and other more 'desirable' species. Indeed, in the Netherlands an enormous machine that dwarfs combine harvesters has been created specifically to 'skim off' the soil surface bearing *Molinia* swards in desperate attempts to keep heathland alive. The idea is that heather will seed into the newly exposed

Flowers (left) *and seedpods* (right) *of yellow flag iris. Both are slightly smaller than natural size.*

subsoil, after the top layer with its extra nutrients has been removed. Whether natterjack toads or other heathland rarities making the most of *Molinia* appreciate the long-term benefits of being processed through one of these nightmare contraptions seems questionable, and there is surely more to be said for preventing the disease (i.e. controlling air pollution) than for this spectacularly damaging type of cure. Other distinctive inhabitants of these impoverished places are the sphagnum mosses, sometimes associated with insect-eating sundews and other plants typical of wet heathland.

Arguably more attractive are the flowering plants of the pond edge. Common among these are the yellow flag iris (*Iris pseudacorus*), the King cup or marsh marigold (*Caltha palustris*) and the delightful small blooms of the water forget-me-not (*Myosotis scorpioides*) and brooklime (*Veronica beccabunga*), but there are many more. Arrowhead (*Sagittaria sagittifolia*), with its distinctive leaves and small white flowers, is widespread in ditches and slow-flowing rivers of southern England. This plant has three different types of leaf: a threadlike underwater one, a flat floater and the barbed emergent variety that gave rise to its popular name. In fact having a variety of leaf shapes is a common feature of water plants, especially the combination of flat floaters with a large surface area (presumably to catch as much sun as possible) and thin, spiky underwater leaves that give minimal resistance to water flow. Then there are the water plantains, with members from several different families, that form yet another distinctive group of marginal plants. Water mint (*Mentha aquatica*) is as unmistakeable as its name suggests, with a scent identical to the garden variety when its leaves are crushed, while the flowering rush (*Butomus umbellatus*), with its pink rosettes, is a singularly attractive waterside inhabitant. Dune slacks, those most temporary of ponds, can be treasure troves to the botanically inclined pondhunter. Marsh pennywort (*Hydrocotyle vulgaris*) commonly prostrates itself across the pond basin, while exotics such as the marsh helleborine (*Epipactis palustris*) and other orchids reward the careful searcher. Pond marginal plants have great variety, and, as with all things, the more you look the more you will find. St John's worts, asphodels, water dropworts and so on will need a bit more skill, a bit more familiarity with the water's edge, and above all, of course, the appropriate identification guide to get you going.

Arrowhead (left) *about half natural size and kingcup* (right) *about natural size.*

What going to the bog really means

It may not seem like it at the time, but getting a bootful of bogwater is different from falling in a fen. Although most people use these terms more or less indiscriminately, there are in fact defined, biological differences that lie behind them. Not only that, but getting wet feet (or not) is actually a serious part of these definitions. So if you want to impress your friends, or just get the crossword right next time, here they are:

A **bog** is a wet area that normally has no standing water (i.e. no obvious layer of water above the ground surface) except in the middle of winter or after unseasonably heavy rains. Moreover, it forms on nutrient-poor land and is usually dominated by mosses (such as sphagnum) and other plants that thrive in impoverished, acidic habitats. Bogs occur on lowland heaths but more often in upland regions; the so-called 'raised bogs' of Scotland, Wales and Ireland are particularly interesting and important. You should be able to cross all of them in summer without water covering your feet.

A **fen** is an altogether richer wetland, otherwise similar to a bog but with water full of nutrients and in which a much greater variety of plants and animals can live. Lagoons, often called 'meres', are or were a common feature of fens. Most famous were the Fens of Cambridgeshire, now largely drained, but in biological terms the wetlands of the Somerset Levels are also fen country. Again, feet should stay pretty dry most of the time.

A **marsh** is a more general term which is usually applied these days to low-lying areas of land with rich water supplies, overlying a mineral (rather than peaty) substrate and intersected by drainage ditches. Famous examples occur around the east and south coasts including Halvergate (Norfolk), the North Kent marshes, Romney Marsh and the Pevensey Levels in Sussex. Again mostly OK on the feet.

A **swamp** is a wetland in which waters covers the ground surface for most of the year, but is not deep or open enough to be called a pond or lake. Feet are in trouble on this one. Extensive areas of reedswamp occur in many places, particularly at some of the Norfolk Broads and at sites like the RSPB bird reserve at Minsmere in Suffolk.

So now you know.

Water plants: the floaters

Hydroponics, the nurseryman's art of growing plants in nutrient solutions without the need for soil to root in, was discovered aeons ago by some of our commonest pond inhabitants. Several water plants grow while simply floating on the surface film, with roots trailing free in the water below, but in this section I also include those that root firmly in the bottom mud but have leaves only on the water surface. There seems a lot of evolutionary sense in having your leaves on top of the water, otherwise you run the risk of being smothered by competitors that cover the surface film and cut off your essential light supply. As usual, though, it's not a perfect strategy for all circumstances. You can get pretty hot in summer, and freeze in winter, if you are stuck at the surface and of course there's always the risk of being washed away if you are not anchored to something. The winter problem is usually solved by the simple expedient of sinking to the bottom, or just dying back to winter roots or tubers in the mud, but the other difficulties are less easily solved and presumably this is why only a small proportion of water plants actually live this way.

Without doubt the best known and most abundant of the true floaters are the duckweeds. There are five species native to Britain, of which the lesser variety (*Lemna minor*) is the one usually seen forming the bright green carpets that so often indicate a stinking pool rich in decaying organic matter. Great duckweed (*Spirodela polyrhiza*) has significantly bigger leaves, as its name suggests, and these are usually purple underneath making identification easy. It's much rarer than its lesser relative, but locally abundant especially in ditches in parts of southern England. Another duckweed of particular interest is the ivy-leaved sort (*Lemna trisulca*), again locally common in the south. This species, which is often mixed with one or both of the other two, generally signifies a richer and less odorous habitat and always attracts my attention as a probable good hunting ground for aquatic animals of many kinds. Finally, because it's Europe's (some say the world's) smallest flowering plant, rootless duckweed (*Wolffia arrhiza*) also warrants a mention. This tiny floater is very restricted in its distribution; its single leaves are less than 1 mm across and, of course, there are no roots hanging underneath. At first sight it's easily mistaken for a floating scum of green algae, so, despite its claim to fame, don't expect a floral extravaganza from this minuscule member of the pond flora.

Other attractive little plants of the water surface are frog-bit (*Hydrocharis morsus-ranae*) and water fern (*Azolla filiculoides*). Frog-bit looks like a miniature water lily, but, like the duckweeds (with which it is often found), has roots that trail freely in the water beneath. Water fern is an introduction from North America that has spread widely in southern England, but tends to be sporadic in its outbreaks. When it does get going it can be dramatically successful, and since it turns red in autumn it then transforms the water surface into a brilliant spectacle quite unlike anything our native plants can manage.

Our largest truly floating weed is confined to rich waters, particularly those with a good supply of chalk which often forms a crusty deposit on its leaves: water soldier (*Stratiotes aloides*) has firm, tooth-edged leaves that can give a nasty scratch if handled carelessly but is a popular home for many water creatures, including crested newts which somehow manage to lay their eggs on it. Like most of the other floaters, *Stratiotes* goes berserk given half a chance and can completely choke up disused canals or suitable ditches. The plants rise to the surface in spring, and display their small white flowers in May or early June.

Of the plants with proper roots or tubers, but only floating leaves, the best known must be the water lilies. There are two common wild species in Britain, one with white flowers (*Nymphaea alba*) and the other, commoner still, with yellow ones (*Nuphar lutea*). *Nuphar* is a very adaptable plant and even grows occasionally in those most acid and inhospitable ponds of heaths and moors. There is also a miniature version, the fringed water lily (*Nymphoides peltata*), which has yellow flowers and

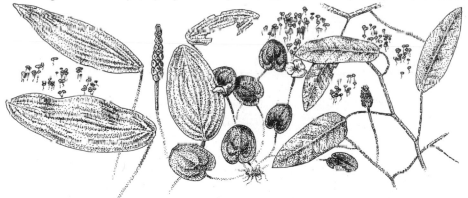

A mixture of floating-leaved plants, all about half natural size: broad-leaved pondweed (left), frogbit (middle) and amphibious bistort (right), all with lesser duckweed mixed in.

slightly serrated leaves and seems to be on the increase in southern England. After the lilies comes a widespread if rather drab equivalent that has no bright flowers at all: the broad-leaved pondweed (*Potamogeton natans*) and its immediate relatives. *Potamogeton* is commoner and more widespread than the lilies, especially in small ponds, and gives much the same impression by covering large areas of the surface with its flat, ovoid leaves from spring through to late autumn. There are of course other weeds with floating leaves apart from these major players on the pond scene, but most are much less frequently met with. One you might well see though is amphibious bistort (*Polygonum amphibium*). This interesting plant can exist in two quite different forms, one growing in marshy places and the other submerged in water with elongate, floating leaves and a spiky, pink emergent flowerhead.

In many ponds, it's the floating plants that dominate and provide that all-important striking first impression. Sometimes they even conceal the very existence of a pond: a duckweed carpet looks all too like a flat, easy short-cut when you come across it in the middle of a wood and the first impression this makes when sinking up to the knees in stinking mud is striking indeed.

Water plants: the real submersibles

What I imagine most people think of as a real water plant is one of those species that grows 'properly' underwater, with most of its roots and leaves (and sometimes even its flowers) never breaking the surface. The typical plant of this class has a feeble stem, needing little strength to support itself underwater, and usually masses of small leaves rather than a few big ones. There is an enormous variety, and space permits mention only of those most often encountered, or sometimes of the most striking. Even this is a bit tricky, since a good few are widely distributed throughout the British Isles and what follows isn't in any order of overall abundance.

It is perhaps unfortunate to start with a weed that is not a native of Britain, but the fact is that Canadian pondweed (*Elodea canadensis*) and its various cultured relatives are now common in waterways all over the country. The original wild type, with short and fairly straight leaves, spread rapidly after its introduction in 1842; this form and a curly variety, with longer, twisted leaves, can now be found in ponds, ditches and canals anywhere and everywhere. At least these invaders haven't fulfilled one prophecy, that they would become so rampant as to exterminate native plants, but they are certainly here to stay and are still the commonest weed in most garden ponds as well as being widely sold in aquarists shops. The lesson hasn't been learnt either, because in recent decades another and perhaps even more invidious foreign plant, the New Zealand stonecrop, has arrived on the scene (see Strangers on the shore).

Two plants sometimes confused at first glance, but really quite different, are starworts (*Callitriche* species) and water crowfoots (*Ranunculus* species). Both have masses of small green leaves, but these have distinctive shapes and only *Ranunculus* has yellow and white flowers that stand proud of the water surface. Starworts, especially the common one of our two native types (*C. pallustris*), are found in ponds and ditches almost everywhere (except my garden ones, where I've never been able to establish it!). Crowfoots are a bit less widespread but often occur as glorious flowering bouquets on suitable ditches, canals and even dewponds. Some species of *Ranunculus* (there are thirteen altogether) are pioneers, colonizing new ponds but disappearing after a year or two; others are the dominant inhabitants of fast flowing chalk streams and

Submerged plants: (from left to right) *starwort, hornwort, Candian pondweed* (straddling right across) *and water milfoil. All about life size.*

similar clearwater rivers.

Hornwort (*Ceratophyllum demersum*) is arguably the most truly aquatic of all our submerged plants; it needs no roots, and drifts around as free-floating masses that can completely clog up those ponds and ditches well suited to its rampant growth. Watercress (*Rorippa nasturtium-aquaticum*) is of course the basis of another pond life industry, but, like gold, has its idiot's equivalent; fool's watercress (*Apium nodiflorum*) is just as common as, and often grows with, the real thing. Leaves of fool's cress are finely toothed, unlike those of the real McCoy which are smooth edged, and there are differences between the flowers too. Water milfoils (*Myriophyllum* species, three in all) have feathery leaves and small flowers rising in spikes above the water surface, while water violet (*Hottonia palustris*) is a plant of southern Britain with splendid swathes of pale lilac flowers raised well above the water in May and June.

Next in line are the pondweeds, which you may well think we have already been talking about, but this name is in fact quite specifically given to a series of plants in the Potamogetonaceae family. Unlike *Potamogeton natans*, which, as we saw in the previous section, is a kind of poor man's water lily, the other members of the group are mostly true submersibles. Many are rare, or at least local in their distributions, but two that you may well meet are curly pondweed (*P. crispus*), also known as frog's lettuce, and opposite pondweed (*Groenlandia densa*). Curly pondweed has reddish brown leaves, very curly and crimped, alternating from one side to the other along the stem, while opposite pondweed has green leaves which, as the name suggests, grow out opposite each other along the stalk.

I will finish with a more or less miscellaneous assortment of underwater plants that you might come across from time to time, though they often live in rather specialized places. Water parsnip (*Berula erecta*), mainly a plant of the pond edge, quite often grows in shallow water and is easily mistaken for fool's watercress. Its leaves are however more coarsely and sharply toothed. Marestail and water horsetails are distinctive and superficially similar plants that are often confused; marestail (*Hippuris vulgaris*) has soft leaves, tiny pink flowers and no terminal 'cones', but water horsetail (*Equisetum fluviatile*), with the opposite characteristics (and no obvious flowers), is much the commoner plant of this pair. In fact horsetails are in the fern family, whereas marestail is a true flowering plant and therefore not at all closely related.

Willow moss (*Fontinalis antipyretica*) is one of very few truly aquatic mosses, often found attached to stones in water with at least some flow to it. Apparently it was once packed around chimneys to prevent roofs catching fire, thus explaining its bizarre Latin title *antipyretica*. Marsh St John's wort (*Hypericum elodes*) covers its chosen ponds, usually rather mineral-poor ones that aren't too acid, with attractive yellow flowers from June to September. The bladderworts (*Utricularia*, three species) are rare but fascinating plants; yellow flowers above the surface bely danger beneath, because these are insect-eaters, and those small bladders strung out along the stems are death traps for mosquito larvae and others that dare to venture in. Like most insectivorous plants, bladderworts are

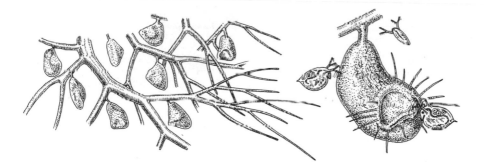

Bladderwort, with enlarged bladder (right) *catching* Daphnia.

inhabitants of nutrient-poor bog and heath ponds where meat makes a valuable addition to an otherwise scant diet.

There are, altogether, literally hundreds of plant species to be found in or around our ponds and ditches. Most are smaller, rarer or both than those mentioned above but it's important to repeat that this is just a flavour of waterside botany. A useful thing about plants is that they can't swim or run away, so you have all the time in the world to make a study if this aspect of pond life becomes your particular forte.

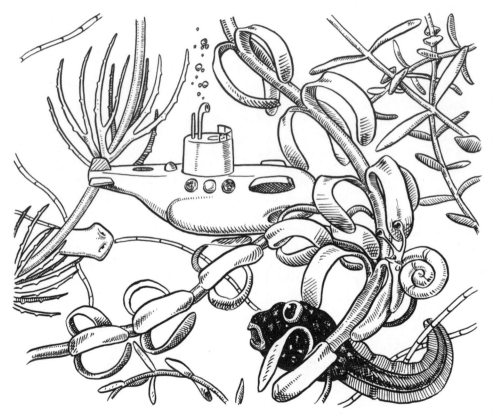

Some true submersibles.

What about the algae?

The previous three sections have dwelt on higher plants, for the simple enough reason that these are the kinds usually seen when you walk up to a pond. Higher plants are, however, far from the whole story: algae, those most primitive greens, are also out and about but in many cases will not be obvious because of their very small sizes (see Microbes). There are some situations, though, where the presence of algae is evident without pouring over a microscope. The simplest case is the bright green pond: a green not caused by duckweed, or any other recognizable plant, but one that seems to be the colour of the water itself. The colour turns out to be due to tiny, individually invisible algae multiplying at a prodigious rate and taking over the whole water system. Algal 'blooms' of this kind usually mean trouble, notably pollution from too much organic waste, phosphate-rich fertilizer or the like. Such superabundance of one or a few micro-organisms usually means that not much else can survive in the water, which becomes low in dissolved oxygen at night when the algae use it up rather than produce it by photosynthesis. Duckponds with too many ducks are all too often in this state, an altogether sad fate for a good few village ponds in recent years.

Another manifestation of algae, this time of a filamentous type, is blanket weed. Mats of this stringy green fibre can indicate pollution again (especially if it's the slimy variety), or just too much sun in the absence of effective competition from higher plants. The latter situation, which is common for example after hot summers in many kinds of ponds, is not necessarily unhealthy; it just looks nasty. Usually a cold spell will make this type of blanket weed recede, and allow higher plants to dominate once again to give the pond an altogether more pleasant appearance.

Finally, some algae aren't that simple after all. Stoneworts (*Chara* species) look very much like higher plants at first sight, with stems and whorls that closely resemble slender leaves. In fact these growths are of a complex algae, but you would have to be an anatomist to prove it. *Chara* is quite common in Britain, though usually its time in a particular pond is limited, since the true higher plants tend to overwhelm it as the years go by. *Chara* is the nearest freshwater gets to big-time algae, but it's salutary to remember that some of the largest plants in the world – seaweeds hundreds of metres long – are members of this 'primitive' group.

Ponds as archives of past disasters

An intriguing recent discovery really makes the best of mud, which after all isn't very attractive for anything else. Taking a core sample through the bottom of an old pond is to travel back in time; obviously the mud on top is the newest and that at the bottom the oldest, and, so long as it hasn't been disturbed too much, the age at various depths can be measured using radioactive decay (much the same way as carbon-14 decay is used to age old wood, Egyptian mummies and so on). Knowing how old a bit of mud is doesn't sound too exciting in itself, though it can tell you how long the pond has existed; but if the remains of micro-organisms in the mud (especially diatoms, those algae with the tough cell walls) are examined as well, lots of interesting insights can be gleaned. Each type of diatom grows best in water of a particular quality, so if the diatom species ('microfossils') change through the mud core it indicates that the pond's water quality has changed as well. This painstaking analysis (imagine counting and identifying thousands of diatoms from every sample under the microscope) has been vital in showing how ponds and lakes in many countries were devastated by acid rain after the Industrial Revolution of a hundred and more years ago; also in the cores you can find recent increases in heavy metals (such as zinc and lead) and even soot particles that can be attributed to power stations burning particular types of coal! All this has provided a dramatic illustration of how pure research, in this case on diatom populations, can be of great value in the 'real' world; the evidence these studies came up with was crucial in persuading governments that acid rain is a serious problem in need of serious attention.

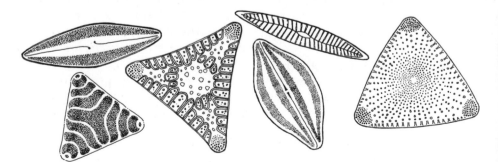

An assortment of freshwater diatoms, 200 times magnified.

Pond animals: an overview

Pond animals can be considered in three main groups, distinguished by convenience as much as by serious biological principles. First come the vertebrates, or backboned animals. These in turn have five orders: fish, amphibians, reptiles, birds and mammals. The first three are known as 'cold blooded' and the last two as 'warm blooded'. Members of all these orders live in or around ponds, though some pools are without any at all, and I put them first because they have the most sophisticated nervous systems of the animal world. What this means in practice is that these are the creatures most likely to see you before you see them; they require the greatest stealth for observation, and even then you may catch no more than the swirl of a fish hightailing it into deeper water, or the crashing of a deer scrambling away through the undergrowth. With care and experience you should do better, but luck is always handy in encounters with vertebrates.

Next in line come the arthropods, which literally means 'joint-legged' animals. They include all the insects, spiders and crustaceans and again there are representatives of all these varieties lurking in our countryside water holes. They are generally easier to approach than vertebrates, but even so their vigilance shouldn't be underestimated. Dragonflies can be tricky to approach closely, and great diving beetles do just that if they catch a glimpse of anything large and unpredictable (such as a pondhunter) wandering about the place. Arthropods include many, if not most of the prime targets for many people interested in pond life and precious few pools are without some.

Finally come the rest, lumped together more for convenience than for sound biological reasons, although it's true to say they are generally the most primitive species. This group includes the molluscs (water snails and mussels), the worms (including leeches), and the true primitives like sponges and hydroids (relatives of jellyfish, but much smaller). They are less complex in structure than the other two groups, and usually combine feeble nervous systems with ponderously slow movement. No trouble catching these, except perhaps for some of the deep-water specialists.

Of course this leaves out all the tiny but vital micro-organisms that live in ponds, but these aren't really animals and will be dealt with separately in a later section.

'Cold-blooded' vertebrates: the lower orders

Fish, amphibians and reptiles share one important attribute: they can't regulate their own body temperatures, but leave the sun to do the job for them. Cold-blooded is therefore a rather silly title, since in summer many of these animals get quite warm, and these days biologists usually call them ectotherms. Whatever title you prefer, many of our ponds have a good variety of such beasties.

Of the fish, the sticklebacks – tiddlers of childhood pond netting – are surely the archetypes. There are in fact two quite different varieties: the 3-spined (*Gasterosteus aculeatus*) and the 9- or 10-spined (*Pungitius pungitius*). It's the 3-spined that most people know, because it is much the commoner and males have the distinctive red chin during the spring mating season. Red is a real turn-on for these aggressive fish, and just sticking a red pencil into a stickleback pond is likely to provoke a vicious attack from a male under the impression that his territory has been invaded by a competitor. Some years back, scientists studying sticklebacks

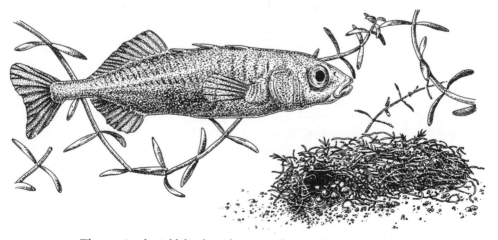

Three-spined stickleback with nest and eggs, about twice life size.

in a laboratory aquarium noticed that at a certain time of day all the males dashed down to one end of the tank; it took a while to realize that this time coincided with the passing outside of the local (red) post office van!

Nine-spined stickles are locally common, especially in ditches in parts of southern England. Males of this species turn jet black at breeding time, with one distinctive white spine on each side. Both species are unusual in that males build nests, with bits of weed and debris, into which they try to entice females for spawning. Three-spined stickles build their nests on the pond floor, whereas those of the 9-spined are usually perched high up in pond vegetation. After the eggs are laid, the males shed their sperm over them and then guard their developing progeny until hatching and beyond. It's one of the few instances of parental care in fish, and equally unusual is the fact that it's always the male that gets the job. These tiny fish are amazingly brave, and I have watched them chase away great crested newts two or three times their size that have dared venture too close to a guarded nest. Stickleback babes grow quickly, and will breed themselves (and probably die) the year after they are born. The spines, of course, are there for protection; arrayed along the fish's back, they literally stick in the throat of many would-be predators. Even so, they have plenty of enemies; water stick insects seem especially partial to them. Sticklebacks are also major predators themselves, taking water fleas and other small items; they can have devastating effects on newts, seeking out and devouring their newly hatched larvae so efficiently as to exterminate them within a few years.

Other species of fish are much less often encountered in small ponds, though ditches are a better bet, especially when these form a large network of intersecting waterways. Roach (*Rutilis rutilis*) and rudd (*Scardinus erythrophthalmus*), with their distinctive orange fins, crop up from time to time and so do carp (especially *Carassius carassius*), tench (*Tinca tinca*), perch (*Perca fluviatilis*) and even pike (*Esox lucius*). I have seen roach spawning on water forget-me-not in quite small ponds, watched shoals of small perch cruising among toad tadpoles in similarly unlikely venues, and caught pikelets that can have been no more than a few days old in the most overgrown of ditches in Norfolk. Of course any tendency to dry up, even if only very rarely, spells doom for all types of fish and this is probably why there are not more species in the small pools beloved of the pondhunter.

In large ditch systems we can safely add bream (*Abramis brama*) and eels (*Anguilla anguilla*) to the list of common encounters, and the latter in particular regularly get caught by vigorous pond netting. The whole business is further complicated by organized angling, since clubs often stock even quite small ponds with coarse fish these days, and by the release of 'pets' such as goldfish (*Carassius auratus*) that frequently survive and breed. Watching freshwater fish, rather than catching them, is a much underrated pastime and can be very relaxing if you find a good spot. However, it has to be said that a prolific fishpond is unlikely to be rich in smaller forms of animal life, for the obvious reason that the fish will usually have eaten them (all native freshwater fish are either carnivorous or omnivorous). Indeed, abundance of amphibian, insect and other small life forms is usually an indication that anything bigger than a stickleback is rare or absent.

Whereas most freshwater fish are the denizens of lakes and rivers, our six native amphibians are all very much pond dwellers during their tadpole life and as breeding adults. Frogspawn, after all, is a pond speciality; the two things go together like witches and broomsticks in the folklore of our countryside. Frogs (*Rana temporaria*) went through a bad patch after the last war but, largely because of the recent vogue for gardens ponds, seem to be getting commoner again now. Harbingers of early spring, the adults usually congregate in February or March and can be watched, in their bustling activity at the spawn site, from quite close quarters if approached with due stealth. Males hold onto females very tightly, and stay put until the eggs are laid; this is usually done at night, and I only saw it in daylight after twenty years of trying. The whole operation takes less than 10 seconds, so you have to be on the ball. Later in the spring, the adults disperse. Frogs are very unfussed about the type of pond they use, and can crop up almost anywhere, including flooded fields and cart-ruts.

Common toads (*Bufo bufo*) arrive a little later in spring and are rather fussier, preferring the deeper ponds (usually at least a metre or so), but it's not uncommon to find places used by both these species. Toadspawn is laid in strings around submerged water plants and is less obvious than the clumps, often laid in large masses together, of the common frog. Again it's quite easy to watch breeding toads, and there are often spectacular fights with males croaking and wrestling for possession of females. Spawning by toads takes many hours to complete, and pairs doing it are not hard to find, but obviously should not be disturbed.

For both frogs and toads, time spent breeding is short, and you will have to arrive during the right week to see the peak of activity. Tadpoles, however, will be around until midsummer, and, apart from securing the next amphibian generation, provide a food source to an enormous

Common frog (above) *and common toad* (below), *just slightly larger than life.*

number of insects and other predators. Probably for this reason, frog tadpoles get very secretive as they grow, hiding a lot of the time in weed or mud; but common toad tadpoles taste as nasty as their parents, and obviously know it, because they swim much more conspicuously in open water and often form large shoals. It's easy to tell the two types apart: frog taddies are brown with gold speckles, whereas toad taddies are jet black. Tadpoles mainly eat debris on the pond floor, but meat is much appreciated when available and this can include dead parents or sickly brothers and sisters. The first two months or so of a tadpole's life are spent just getting bigger, but by the end of May hind legs make an appearance and, in normal weather conditions, front legs emerge two or three weeks later at the start of the big change into frog or toad that we call metamorphosis. If all goes well, the pond banks in June or July will seethe with froglets or toadlets for a week or two before they disperse to drier places and new hunting grounds. It will be two or three years at least before the survivors return to the pond as breeding adults, and a lucky toad may live fifteen years or more though most die before they are half that age. For frogs, the news is bleaker: without the poisonous skin of their warty relatives, they taste too good, and a real survivor will be no more than seven or eight.

Common frogs and toads are found throughout the land but our third species of tailless amphibian, the natterjack toad (*Bufo calamita*), is much rarer. With a preference for temporary pools, natterjacks are most often found on the coastal sand dunes of north-west England and are easily distinguished by their yellow back-stripes. Males also call very loudly during their later and longer breeding season, and are much noisier than common frogs or toads. It's one of our strictly protected species, and if encountered should be left very much alone. Natterjack spawn is similar to that of common toads, but is often laid in more open places like on a flat, sandy pond bottom rather than twined around plants and has a single rather than a double row of eggs. Tadpoles look almost exactly the same as common toad ones, but usually develop a white 'chin patch' when about half-grown that never appears in common toad taddies. Because natterjack tadpoles grow so quickly they too metamorphose around mid-June, and the yellow back-stripe makes its appearance at this time so identification is no longer a problem.

Our remaining three amphibians, the newts, again for many of us evoke memories of childhood and the wielding of pond nets. Smooth or common newts (*Triturus vulgaris*), palmate newts (*T. helveticus*) and great crested newts (*T. cristatus*) are all widespread in Britain and it's still possible to find ponds with all three together (sometimes even with common frogs and toads as well, a veritable amphibian Eden). All three species journey to their breeding ponds in early spring, and stay there until early summer, so it is easier to come across them than it is to meet up with the short-stay frogs and toads. It's also easy, and delightful, to watch their elaborate courtship displays. Males chase females beneath the water, bounding in front and fanning their tails furiously to attract attention. A receptive female will stay still, and later on pick up a package of sperm dropped by the male on the pond floor. In the ensuing weeks she will lay her few hundred eggs, one at a time, wrapped in the leaves of pond plants. Knowing this, it's quite straightforward to detect the presence of newts in a pond just by searching for these distinctively folded leaves during April and May. Newt tadpoles look like miniature versions of their parents, but with large feathery gills at neck level, and hide most of the time in thick weed. They develop feeble-looking front legs first, soon after birth, and the back ones follow a while later so newt tadpoles have all four legs much earlier than their frog or toad relatives. Newts are predatory at all ages; adults eat anything they can swallow and are pathetically bad at judging what is an appropriate meal. It's commonplace to see newts grab worms much larger than themselves and wrestle for ages before it dawns on them that their task is impossible, but real food items include *Daphnia*, water hog lice and other manageable invertebrates (see p.79). Newt tadpoles seek out tiny crustaceans too, and in particular share their parents' delight in Daphnia. They in turn are eaten by a variety of predators, including all those fierce invertebrates that consume frog and toad tadpoles but with the additional hazard of parents not averse to a bit of cannibalism now and then. Newt tadpoles leave the water much later than frog or toad taddies, usually in August or September but sometimes not until the following spring. Like frogs and toads, they will be back in a couple of years or more to start the cycle again but while away from the pond will live in thick grass or under stones, emerging on damp, mild nights to hunt for worms, slugs and small snails.

Despite all these basic similarities, there are significant differences between our newt species. Palmate newts are the smallest, and the male has webbed hind feet as well as a peculiar, truncated tail. They do especially well in nutrient-poor ponds on heaths and moors, where they can be very abundant and not infrequently are the only newt species present. Smooth newts are just a little larger, and the male has a well developed, wavy crest at breeding time. They are found almost everywhere except in the most acid of heathland ponds. Female palmate and smooth newts are difficult to tell apart, though the former usually have pinkish unspotted throats whereas female smooths have whiter throats normally with some black spots. Great crested newts pose no identification problems; they are much bigger than the other two, look black or at least very dark from above, and have bright yellow or orange bellies with black markings. Males, as the name suggests, have magnificent, toothy crests in spring. Although much commoner than natterjack toads, crested newts are also strictly protected under the law. It's fine to watch them in a pond, but don't deliberately catch them, and release any caught by accident as quickly as possible. Crested newts often crop up in rich, weedy ponds but rarely with fish. Their tadpoles are

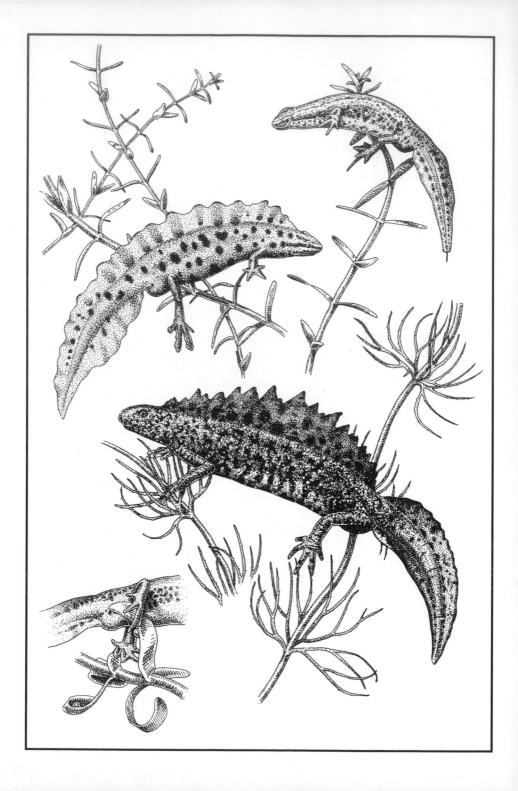

especially prone to being eaten by fish, because they like to swim conspicuously in open water.

All our amphibians, be they frogs, toads or newts, have to spend the winter sheltered from frost in a state of hibernation. Most find themselves a hidy-hole on land; piles of dead leaves, compost heaps, drystone walls or simple cracks in the ground are favourite places. Some of all the newt species, however, as well as a proportion of most frog populations, choose to hibernate underwater in the bottom mud of a favourite pond. Presumably this is safer from predators than land-based sites, but it can be a dangerous option in severe winters and, when ice covers a pond for long periods, it's not uncommon to find a few dead amphibians (suffocated because oxygen can't get in through the ice) floating about afterwards.

Last but by no means least, spare a thought for pond reptiles. Britain is not too well off in this regard, without any of the terrapins or crocodilians of warmer climes; but we do have the grass snake (*Natrix natrix*). This, alone among our native reptiles, is a regular inhabitant of watery places. It's also our largest snake, frequently reaching 90 cm (36 in) in length and with individuals twice that size on the record books. Its love of water is strictly of the cupboard variety: amphibians, particularly frogs, are the grass snake's main food and it is happy to swim and dive in the deepest pond to search them out. Watching snakes winding their way across the water, heads held high above the surface, is a striking and unforgettable sight rather reminiscent of the cruder Loch-Ness monster hoaxes. I've caught small grass snakes in pond nets without ever seeing them first, almost certainly in the act of chasing tadpoles. Grass snakes are commonest in southern Britain, and are most often met with on undisturbed heathlands or the marshy lowlands of some river valleys.

Newts: male palmate (top, right), *male smooth* (top, left) *and male crested* (centrepiece); *bottom left shows female laying eggs in leaves. All about natural size.*

'Warm-blooded' vertebrates: the upper echelons

Considering that mammals and birds are the evolutionary sophisticates of the animal world, rather few of them grace our pond sides with their presence. In fact there are only two mammals that can be considered as pond specialists: the water shrew (*Neomys fodiens*) and the water vole (*Arvicola terrestris*).

The water shrew is the largest of Britain's three shrew species, but that isn't saying much; 9 cm (3½ in) is the most you can hope for, excluding the tail, which is as long again. It is quite a common animal, living in holes in the banks of ponds and ditches where it hunts underwater both by day and by night. Despite that, and rather poor eyesight, it's not often seen. Black-and-white fur on land turns into a ball of silver, due to the trapped air, when diving underwater. Shrews are strictly carnivorous and this species hunts tadpoles, insect larvae and any other meat that happens to be available. They're so good at it I often wonder why there aren't more about.

Around May or June, the female makes a comfortable nest out of moss, grass or dead leaves deep within her burrow and shortly afterwards gives birth to 5-8 blind, naked young. These grow quickly and are independent less than six weeks later, giving mum a chance for a second brood later in the summer. Shrews don't hibernate and are active all year round, but their lives are brief: 15 months is a good age for these inhabitants of the metabolic fast lane, burning up food and energy at a staggering rate, to expire exhausted in the hungry days of early winter. Apart from this death by natural causes, owls are said to be major water shrew predators and a fair few end up in the stomachs of pike as well.

Water voles, sometimes called water rats, are much bigger and at up to 22 cm (8½ in) long, are easily our largest vole species. With flat face, short tail and grey-brown fur, there's no mistaking them for shrews, or for real rats (*Rattus norvegicus*) that also take to water quite often. If a pond has water voles it will usually have a colony of them, and, since this animal is as active by day as by night, your first clue may well be the splashes as they

Water vole (left) *and water shrew* (right), *both about half natural size.*

plop into the water when you approach. Voles are mainly vegetarian, browsing on water weeds and reeds, but they do also eat insect larvae and I once saw one carrying away an adult frog. Although aquatic, water voles don't like rain and any downpour is sure to keep them at home deep in the burrow. Some of these burrows have underwater entrances rather like beaver lodges, and networks of vole tunnels have irritated Fenlanders by the damage they cause to embankments designed to keep water out of low-lying fields. Like shrews, they don't hibernate, but more shrewdly than shrews they lay up stocks of nuts, acorns and water plants to see them through the lean months of the year. Breeding goes on through the summer, with about two young at a time born in a grass-lined nest deep underground. Like most small mammals, their lives are usually short and dangerous although they generally survive a bit longer than the frenetic shrews that share their watery habitat. The water vole is a species which, although widespread, seems to have declined in recent years and no-one really knows why.

Because of their general abundance these days, brown rats are perhaps the most likely mammals to be seen swimming across ponds or ditches although they are not waterside specialists. Rats are easily identified by their more pointed noses and much longer tails (relative to body size) than those of their vole counterparts. Two other mammals are also worth a mention: North American mink (*Mustela vison*) have escaped from captivity and spread across large areas of Britain in recent decades. Occasionally they visit small ponds, where they prey on fish, water birds and small mammals. Attempts to eradicate them have failed dismally, and mink are becoming increasingly common in many areas. Conversely the native otter (*Lutra lutra*), which is much bigger, has all but disappeared from much of lowland Britain. I have watched otters in large ponds years ago, and in East Anglia they hunt along drainage ditches, but you'd be lucky indeed to see one now and in any case they are more at home in larger lakes or rivers and on wild seashores.

Turning to birds, the difficulty is one of where to draw the line. Innumerable species turn up occasionally on ponds and ditches, but in my experience only a very few are regularly encountered on small water bodies. Foremost among these are moorhens (*Gallinula chloropus*) and mallards (*Anas platyrhynchos*): the former especially crops up in the tiniest and muddiest of pools, diving for cover when humans approach or half running, half flying across the water surface to the safety of an opposite bank. It's rare to find a lot together, but moorhens are very common birds across the whole country. They are omnivores, taking a mixture of whatever plants and animals they can find. Mallards, too, crop up as odd pairs on small ponds and in narrow ditches; usually, though, there's a safe haven such as a reed patch, small island or tree branches dipping into the water where they can roost at night or even nest in spring. Mallards are also omnivorous, and certainly catch animals as large as adult frogs when they get the chance.

Pond life is not really about birdwatching, so I'll do no more than mention a few other pondy species that will be met occasionally. Most of the obvious contenders, such as coots, many of the other ducks, great crested grebes and swans usually prefer lakes or very large ponds, though

Moorhen, about one-third of life size.

A mallard drake.

it's surprising how often mute swans (*Cygnus olor*) make do with quite narrow drainage ditches and I have even seen Canada geese (*Branta canadensis*) take up residence on shallow, temporary ponds used by natterjack toads on heathland. Little grebes (*Tachybaptus ruficollis*) also seem happy to live in smallish pools, and snipe (*Gallinago gallinago*) are regularly flushed from ditch banks by passing pondhunters. And that, in my book, is about it.

Bounders, bugs and beetles

There can be no doubt that if anyone ever writes a complete history of life on earth, and they are fair about it, by far the largest chapter for any one class of animals will go to the insects. In their many and varying forms they get almost everywhere (though, oddly, not the seas and oceans) and ponds are often brimming with them. Freshwater insects can be grouped, again for convenience, into four lots: the springtails (just a couple of aquatic types); the water bugs (fifty or so species); the water beetles (a few hundred species); and the larvae of flying insects, including moths, mayflies, alderflies, stoneflies, dragonflies, 'true' flies (mosquitoes, etc.), caddis flies and some other minority groups (again a few hundred species). Remember that this is convenience, not serious science: caddis flies are no more closely related to mosquitoes than they are to water beetles.

Springtails look at first sight to be no more than tiny black dots, usually bunched in groups that resemble a puff of soot floating on the water surface. Only 2 mm long or less, they can bounce off the water and several centimetres up into the air when using the appendage that gives them their common name. Springtails (of the insect order Collembola) feed mainly on vegetable matter, and are themselves a popular food for small animals such as baby frogs and toads on their way out of the breeding ponds. Springtails can crop up in large numbers if you happen to be in the right place at the right time; like many animals, they have cycles of abundance and rarity, and it's possible to go some time without finding any at all.

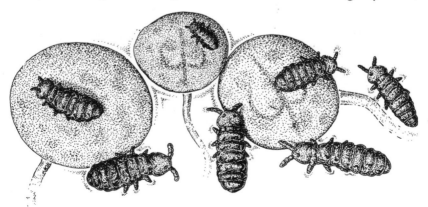

Springtails, ten times life size!

Water bugs have wings (though these are not always obvious) and the distinction of a single, piercing beak rather than pairs of jaws that crunch together. This means that some of them can give you a good stab if not handled carefully, so be warned! Most obvious, however, are the most harmless (at least to people): pond skaters (*Gerris* species), water measurers (*Hydrometra* species) and water crickets (*Velia* species) are all denizens of the surface film and there are several varieties of each in

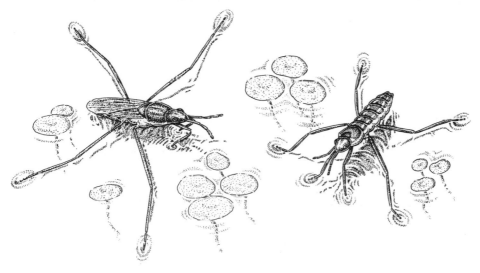

Water boatman (left) *and water cricket* (right), *both twice as large as life.*

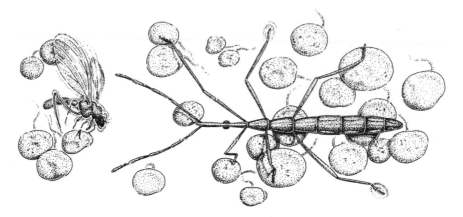

Water measurer (right), *five times life size, making a beeline for a struggling fly.*

Britain. Pond skaters are the largest and commonest, frequently up to 2 cm (¾ in) or so long; they fly strongly, and rapidly colonize new ponds, so it's rare to find water without at least one type skimming effortlessly across the top. Water measurers are infinitely more delicate, smaller and thinner versions that tend to hide up more in bankside vegetation but are nevertheless pretty common. Water crickets are somewhere in between in terms of size and stoutness, but are easily identified by the orange-red marks along their sides. All these beasts rely for survival on the misfortunes of others; they constantly patrol the surface film for small flies unlucky enough to have become stuck in it, which then suffer administration of that piercing beak (which, in these bugs, is too feeble to enter human skin).

Other very noticeable bugs are the water boatmen. Of these there are two main types, each with several representatives in Britain: the greater (*Notonecta* species) and the lesser (mainly *Corixa* species, but also some other close relatives). The differences between greater and lesser are not really to do with size (both grow up to 15 mm or longer) but are nevertheless easy to spot: greater water boatmen swim upside down (they're often called backswimmers) and are naturally buoyant. Common

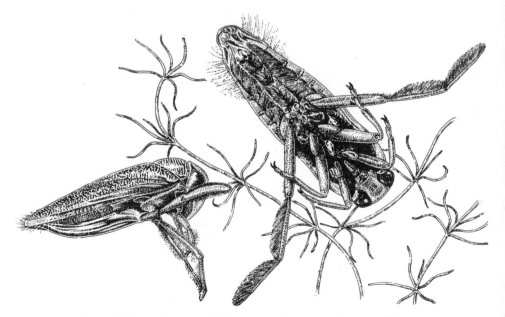

Lesser (left) *and greater* (right) *water boatmen, five times life size.*

and usually obvious inhabitants of open water in ponds everywhere, they stop swimming and rise effortlessly to the surface every so often. Lesser boatmen are the direct opposite; they swim right side up, and have to power themselves upwards to replenish their air supply. None of these bugs has gills, so all must come up at intervals for this purpose. There's another difference that is even more important to the pondhunter: *Notonecta* species are aggressive predators, grabbing insects on the surface film from underneath if they can get there before the pond skaters. They have a positively fiendish beak, which will not be forgotten once experienced. Lesser boatmen, by contrast, sift through the muck on the pond floor using a feeble, hoover-like mouth; they are essentially scavengers, and can be handled in complete safety.

Just three other water bugs are common and notable. The water scorpion (*Nepa cinerea*) looks fearsome, and no doubt is if you happen to be a tadpole, but its beak is weak and I've never been at the sharp end of it. Similar in principle, but longer and much thinner, is the water stick insect (*Ranatra linearis*). This cannot hurt people either, and both lurk in dense weed in wait for other insects, tadpoles or small fish. *Nepa* is by far the more common of the two, but *Ranatra* can be locally abundant in warm

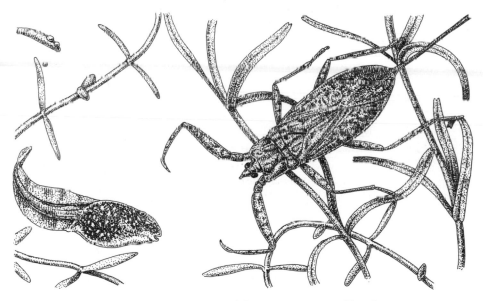

Water scorpion, twice life size, in pursuit of lunch.

Water stick insect, roughly life size.

(sunny) ponds and ditches of southern England. I have come across places where several emerge in every haul of the net, but admittedly not often. Last, but by no means least, is the saucer bug (*Ilyocoris cimicoides*). Twenty millimetres or so long, it frequently occurs in numbers together and at first sight looks very much like a water beetle. Look carefully underneath, though, and you will spot a beak which is, if anything, even more unpleasant than that of *Notonecta*. Another inhabitant of dense weed that preys on any small animal unfortunate enough to get in its way, *Ilyocoris* is locally common in much of the country.

Bugs have no distinctive larvae, and youngsters look much like adults though there are usually minor differences in shape and colour. Most species mate in early spring, and lay their eggs in or on the stems of water weeds. This is cunningly timed so that the hatching youngsters emerge just when food is becoming plentiful, in the form of spring hatches of flies, tadpoles and so on; in May and June ponds often teem with half-grown water boatmen and their relatives cashing in furiously on this bonanza. The survivors overwinter as adults tucked away in the bottom mud, ready to start all over again next year.

Beetles, the next group to be considered, are rather different. Most adult water beetles are super-streamlined for speedy swimming, and, though that means they are pretty useless at getting around on land (legs point in the wrong direction), many are surprisingly good fliers. There are several hundred species of water beetles, and becoming an expert water-beetler can be a lifetime's work. These insects usually mate up in early spring, though some prefer autumn, and pairs can often be seen swimming around together. Males of many species have suckers or similar attachments on their front legs that help them hang on to females; fertilization is internal, and, as with bugs, the eggs are usually laid in or on the leaves of water plants shortly afterwards. Beetles have larvae that look nothing at all like the adults, but are usually easy enough to tell apart from other sorts of larvae, such as those of the various flies, which also live in freshwater. Your average beetle larva has a fierce looking pair of choppers, like miniature scythes, at the front end, and a long, segmented body with just a couple of short protrusions used for taking in air on the tail tip. The only other larvae that look a bit like them, and are found quite often in ponds, are those of alder flies. The fly larvae, however, have feathery gills down each side of the body whereas the great majority of water beetle larvae lack these features. Almost all water beetles and their larvae have to surface periodically for air; adults of a few species come up head-first, but most of the commoner types stick their rear ends up to the surface when they want to breathe. Growth of the larvae is rapid, those hatching from eggs in April being full-size by July or early August. They then crawl out of the ponds, bury themselves in mud and transform first into pupae (which are therefore rarely seen) before crawling out again in September or October as newly formed adults. Water beetles usually overwinter as adults and some species can live several years before dying of old age, providing of course they aren't eaten by some fish or water bird first.

Best known, and much prized by casual pondhunters, are the great diving beetles of the *Dytiscus* family. These are Britain's commonest big water beetles, and few decent-sized ponds are without them, though they swim strongly and take a bit of catching. Commonest of the group is *Dytiscus marginalis*, a handsome insect but a ruthless killer to boot, tackling anything from worms to small fish. Eggs of this insect, which look like creamy-white miniature sausages 5 mm or so long, can often be seen on the leaves of water plants in early spring. The larva, which grows up to 50 mm (2 in) or so, is if anything even more brutal than its parents. There are actually six species of *Dytiscus*, all 25-38 mm (1-1½ in) long as adults; most of them look quite similar to *marginalis* with mainly yellow undersides; one that you might notice is D. *semisulcatus*, which is jet black underneath and also fairly common.

Our only other large water beetle is the great silver, *Hydrophilus piceus*; this magnificent insect is 40 mm (1½ in) or more long, rivalling the stag as

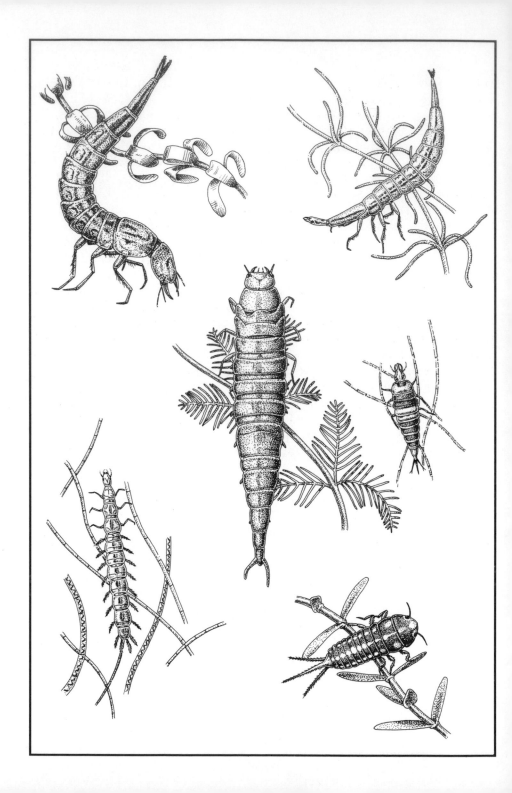

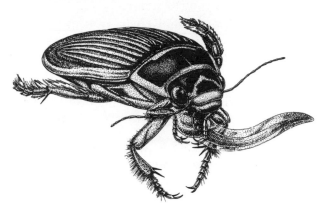

Great diving beetle (above) *and great silver beetle* (below), *both 1½ times life size.*

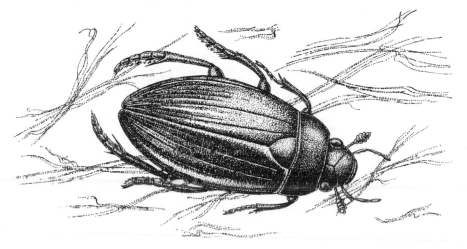

Britain's largest beetle, and from above looks totally black. Seen in an aquarium, however, the reason for its name becomes clear: it carries beneath it a large, silver air bubble which is replenished occasionally by a head-upwards visit to the surface. Unlike the diving beetles, *Hydrophilus* is a vegetarian, browsing on water weed in ditches of south and eastern England; in Victorian times it was a popular aquarium pet, sold in aquarists shops throughout the country. Its larva, however, is carnivorous and grows up to 70 mm (2¾ in) long with enormous can-opener style

A variety of beetle larvae: clockwise from top left: great diving (x1½), sulcated (x2), Hyphydrus (x2), screech (x2) and whirligig (x4), with great silver as centrepiece (x1.25).

jaws that crunch noisily through snail shells en route to its favourite meal. They hatch from silky cocoons, 2 cm (¾ in) or so across and containing about 50 eggs together, carefully spun by the female beetle and left to float on the ditch surface complete with a special 'mast' to provide a reliable air supply. Silver beetles are much rarer than *Dytiscus*, and an exciting find in the net of any pondhunter. All these big beetles have spines or jaws capable of making a lasting impression on the unwary, so take care when handling.

Of the many smaller beetles, a few are worthy of special mention. The sulcated beetle (*Acilius sulcatus*) and lesser diving beetle (*Colymbetes fuscus*) are both brownish insects about half the size of *Dytiscus* (15-18 mm/½-¾ in), and both are common predators in ponds throughout the land. Then there are a lot of small-to-medium size (8-15 mm/⅓-½ in) black beetles without common names, mainly of the *Agabus* and *Ilybius* families, that are probably the commonest water beetles of all. Other notables are the screech beetle (*Hygrobia hermanni*), a small (10 mm) job that announces its presence in the pond net just as its name suggests, and a tiny (3 mm) almost spherical red beetle (*Hyphydrus ovatus*) that scampers around at what seems an enormous speed. Last, but by no means least, are the whirligigs (*Gyrinus* species). These 5mm-ish black bombers have invaded pond skater territory, and on patches of open water can often be seen skitting around in dozens like dodgem cars on the pond surface. They can easily dive beneath, though, and each eye is divided in two: the upper bit to look skyward, the lower to watch for goings on beneath the waves. Not surprisingly, you have to be quick to catch them.

Obviously, with so many beetle species, you will regularly come up with ones that fit none of these descriptions. However, those mentioned above will be the ones most often met and you will need enthusiasm and a specialized key (see the bibliography) to tackle this group in any more detail.

Skating on thin water

One part of a pond worthy of special mention is the surface. Water has peculiar properties where it meets the air, forming the so-called 'surface film' rather like the skin on a rice pudding (but very thin, of course). This skin can't be seen directly, but the effects of it are crucial to the way pond environments work. Most obvious exploiters of it are the insects that buzz about on its surface – the pond skaters, whirligig beetles and so on. Look carefully at the feet of a pond skater and you will see the depressions they make, weighing down on but not penetrating this delicate surface fabric. All these frail-looking creatures rely totally on the existence of the surface film for their mode of life, not just because it supports them and stops them sinking, but also because it provides their grub. Tiny flies and other insects that make up the diet of the film-dwellers learn about surface films the hard way; dropping onto it is for them like landing in glue, and their struggles send ripples across its surface to attract the attention of their hungry predators. Wiggle a blade of grass on the film, and watch them come skating over expecting a juicy meal.

The surface film is important for other reasons too. Its tension is the basis of capillary action, whereby water flows upwards through thin tubes without needing to be pumped. If the tube is thin enough, upwards can mean higher than 9 metres (30 feet)! This is the way plants around the pond extract the life-giving moisture through their tiny stem capillaries, at no cost at all in terms of energy. It's also why covering artificial pond liners with sand or soil so completely that dry soil is continuous around the outside looks fine but can be bad news; capillary action in a hot spell will literally bleed your pond dry.

Recent research has shown that even the chemistry immediately in and around the surface film differs from the proper water below. In particular, lots of nutrients accumulate in the film, which is why you can watch tadpoles or pond snails swimming or crawling upside down and grazing this invisible food supply from beneath. Tasty but dangerous, because it's difficult to imagine a more exposed place to eat and many predators are adept at catching these surface grazers at work.

Fly fishing

An astonishingly large number of flying insects have opted to lay their eggs in water and experience their larval development there. For the pondhunter, this has the practical effect of producing a variety of odd-looking creepers that need identification. In fact it's not too difficult to sort out what general type of fly larva you have encountered but, as with beetles, identification to species level is too daunting a task for most of us.

Commonest of the fly larvae, and probably the least popular, are those of mosquitoes and midges. Cropping up in the grottiest of habitats (water butts are a favourite place), mozzie larvae hang from beneath the surface film and dive down in a kind of twitching motion when disturbed. Pupae have much larger heads, but are equally active; all make excellent fish food, which is probably why they abound in water little else will live in. Mosquitoes seem to be around at almost all times of year, and a mild spell in January will see a new crop of larvae in any available standing water.

Midges are quite different from mozzies, and one of the commonest of the non-biting types, *Chironomous*, has a bright red, segmented larva ('bloodworm') that lives in tubes of bottom mud stuck together with silk and swims with a bizarre, looping motion. Biting midges of the *Chaoborus* family have quite different larvae that hang in the water like miniature, transparent submarines; when disturbed they jerk rapidly away, a habit which has earned them the title of phantoms. Other so-called 'true' flies with aquatic larvae include some craneflies (*Tipulidae*), horse flies (*Tabanidae*) and hover flies (*Syrphidae*). Most of these larvae are

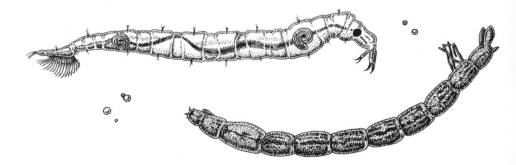

Biting (above) *and non biting* (below) *midge larvae, about five times life size.*

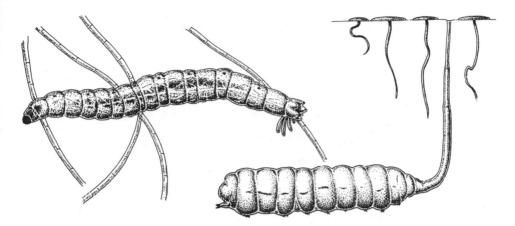

Cranefly larva (top left) *and rat-tailed maggot, both about twice life size.*

segmented, maggot-like crawlers, and one of the best known, that of a hover fly called *Eristalis*, has a long extensible breathing tube and enjoys the popular title of 'rat-tailed maggot'. Delightful. Like the midges and mosquitoes, the larvae of other true flies are rarely found in 'proper' weedy ponds, but prefer marginal habitats such as muddy puddles or pools which support few other animals that might eat them; they can be found in spring and early summer.

Moths and butterflies form one of the few insect groups that steer pretty clear of water, but not entirely so. Look carefully at the leaves of water lilies, or those of other large floating plants, in midsummer: any neat, cut-away sections? If so, you may well find the caterpillar of a china-mark moth lurking beneath in a cell made from the neatly excised leaf segment. The adult moths lay their eggs on the floating leaves, and, on hatching, the young caterpillars mine their way through the plant tissues scoffing the spoils as they go. Later on they emerge to build their cells under lily leaves, holding them together with homespun silk and hibernating there through the winter months. Next summer they will spin a silk cocoon in which to pupate as a chrysalis, the small and unremarkable adult emerging some time after to lay more eggs and die just a few days later.

Superficially similar to moths when seen on the wing, but with characteristic very long antennae and hairy rather than scaly wings, are the caddis flies. Active mainly around dusk on warm summer evenings, the eggs are laid in or above water (into which they later drop) and these quickly hatch into larvae famous for the homes they make, sticking

Do-it-yourself home building.

together bits of gravel, plant leaves or sand grains to create a safe haven they then have to cart around with them. Look in the net for any suspicious, artificial looking structure with a hole in one end. When put back in water and left a few minutes, a head, some body and short legs will emerge and waddle away (case and all) if a caddis larva is in residence. These larvae are mostly vegetarian, but some are predatory and will, for example, catch and eat hatching tadpoles. There are nearly 200 species, but each larva builds a distinctive style of home, so this is one group where identification to species level is often easy. A proviso, though, is that caddis larvae may change plan if their preferred building materials are not to hand and this can occasionally confound reference to the standard texts. Indeed, a popular pastime of naturalists used to be to gently deprive a caddis larva of its case and then provide it with bizarre objects (such as coloured glass beads) which are then used to create a spectacular new abode in the worst possible taste. Caddis larvae spend at least a year in the pond, sometimes two, after which they pupate in their cases (still in the water) before emerging as rather short-lived adult flies. Despite the efforts put into home-making, caddis larvae have numerous successful predators. The larva of one species of great diving beetle (*Dytiscus semisulcatus*) specializes in them; trout relish them; and I have watched crested newts

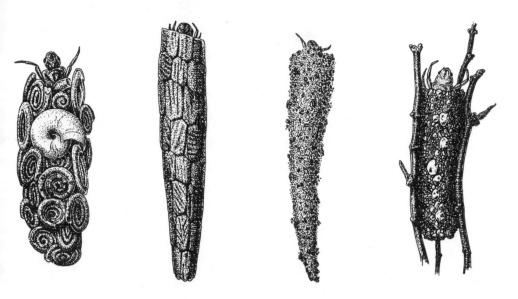

A variety of caddis fly larvae in their cases, all about twice life size.

seize protruding heads, pirouette with spectacular flourish to detach the unfortunate insect's retaining hooks from its case, and gobble up the victim in a matter of seconds. On top of that, some species of ichneumon fly successfully parasitize them; makes you wonder whether case-making is really worth the bother after all.

Alder flies (*Sialis* species) have already been mentioned; these larvae, with jaws, gills curving up from the body segments but no wing pads, are often found under stones, bits of wood or other objects on the pond floor.

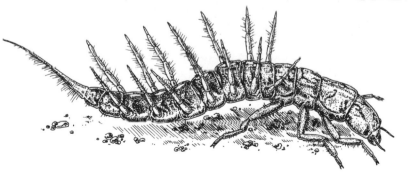

Alder fly larva, four times life size.

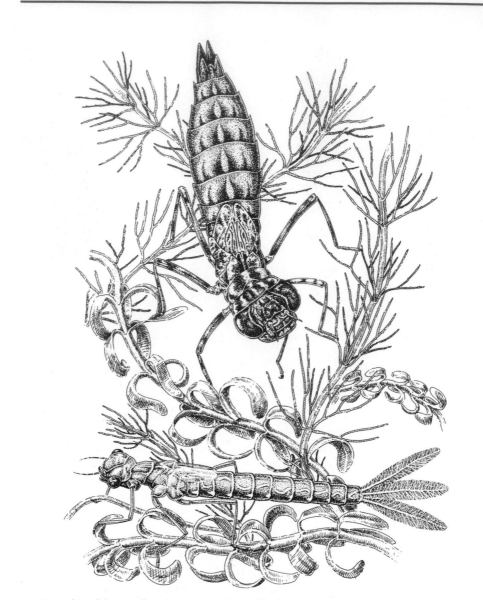

Nymphs of dragonfly (above) *and damselfly* (below)*, two and three times life size respectively.*

Their tails taper to a single point, another useful feature which helps distinguish them from the few beetle larvae that are also equipped with feathery side-gills. Wing pads, looking like miniature wings on a larva's back, are an important clue to identity because the remaining types all have them.

Adult alder flies lay their eggs on foliage overhanging water, and the emerging larva drops conveniently in with the expectation of spending two full years there. After no less than ten skin moults, the full-sized larva crawls out of the water in spring to pupate in damp soil near the pond edge. A few weeks later the adults emerge to complete the cycle.

The damsel and dragonfly family (Odonata), with about forty species, is undoubtedly one of the most important and the larvae, known as nymphs, are easy to tell apart from other groups. The large dragonflies have correspondingly large nymphs; the big 'hawkers', mainly of the *Aeshna* group, have nymphs up to 50 mm (2 in) or so long with stout, elongated bodies, whereas the shorter, fatter 'darters' (including the *Libellula* species) have short fat nymphs to match. These are among the largest insects in the pond, but the nymphs of the more delicate damsel flies are much smaller and frail-looking. None of these larvae has gills down the side of its body, but all have gills somewhere; dragonfly ones are internal, and water with its oxygen supply is sucked in and out via a hole at the

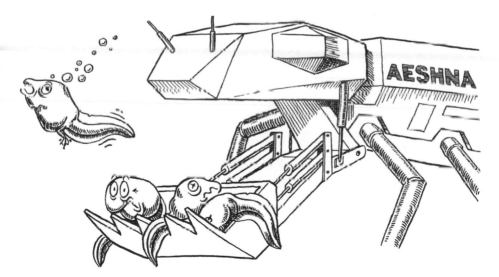

When it comes to the crunch, dragonfly nymphs are made for the job.

back end. This is also very handy in times of emergency, when the water is expelled under pressure and jet-propels the nymph out of harm's way. Damsel nymphs, by contrast, have three feathery gills protruding conspicuously at the rear end. They are very common in most ponds, although telling which species they belong to is not easy. Another feature of all these nymphs is an extensible 'mask', rather like the front end of a JCB digger, that shoots out in front of the head to catch prey.

All these larvae are strictly carnivorous, the large *Aeshna* nymphs catching animals as big as tadpoles or small fish. After one to three years in the pond, the fully grown nymph chooses a warm summer evening to climb laboriously out of the water up the stem of a rush or reed; here the skin splits along the back and, slowly but surely, the adult damsel or dragonfly emerges. The whole process takes several hours, but is well worth watching should you get the chance. The first time I saw it remains an abiding childhood memory, and rather a better one than my first water beetle bite. Adult dragon and damsel flies, surely among the most

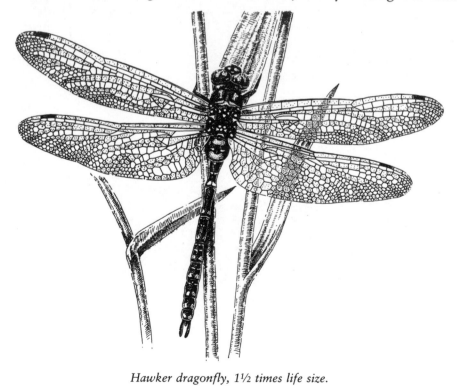

Hawker dragonfly, 1½ times life size.

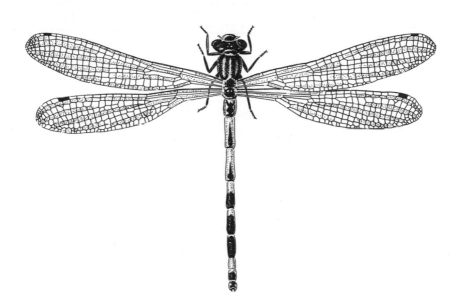

Damselfly, twice natural size.

beautiful of insects, mate on the wing and can often be seen – and heard – flying around in tandem with the female dipping her tail in the water to release her eggs. They're very vulnerable at this time because it's obviously tricky to fly fast *in flagrante delicto*, and I've seen many a pair snapped up by passing birds and leaping marsh frogs. You can also hear them clattering when fighting. For much of the summer dragonflies range far from water, hunting other insects along hedgerows and woodland glades. Sadly, none will survive a British winter and it is left to the nymphs to ensure the next season's generation.

Stonefly larvae must be mentioned in passing, because, although most live in fast-flowing streams, rivers or deep lakes, a few occasionally crop up in ponds and one or two species are even temporary-pond specialists. Their characteristic features are the presence of two, usually quite long tail filaments and often a rather flattened body. The biggest are more than 30 mm (1 in) long, but most come much smaller at 10 mm (¾ in) or so.

The only other larvae that crop up regularly are those of the mayflies. These have superficial similarity to damsel fly nymphs, are of similar size (up to 20 mm/¾ in or so) and are also pretty common. Like the damsels, they have wing pads and three posterior protuberances; unlike them, but like alder fly larvae, they have gills along the sides of their body. These

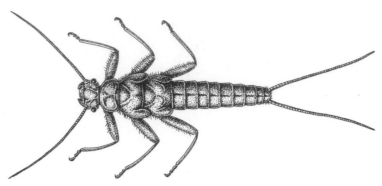

Stonefly nymph, five times life size.

may be feathery and obvious, or plate-like and less evident without careful examination. They hatch from eggs either dropped in the water or, in the case of some species, deposited by adventurous females crawling down weed stems below the surface. Another peculiarity is that, whereas in some species the eggs hatch soon after laying, in others they remain over winter and hatch the following spring leaving only a couple of months for

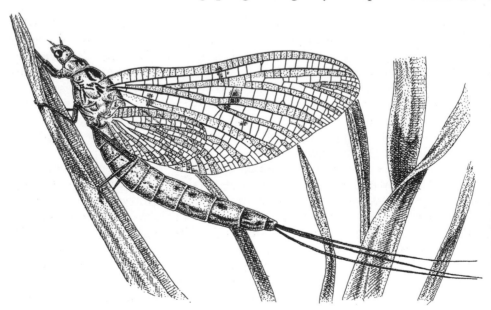

Newly hatched mayfly, four times life size.

the larvae to grow full size. They probably hold some kind of world record for skin changes too, as many as 27 moults being needed by some species. As with damsel and dragonflies, there is no pupal stage, mature nymphs simply rising to the surface when growth is complete. Mayflies metamorphose 'en masse' during late spring, to give the fly hatches of which trout (and trout fishermen) are so fond; for a few evenings the sky above the water hums with these dainty insects, but after mating they quickly die and won't be seen again for another full year. Indeed, becoming an adult in the fly world all too often means that life is almost done. Even the magnificent dragons get just one summer on the wing; autumn's first frosts will see the end of them.

What's in the net? A quick guide to those funny-looking larvae

Beetles, bugs, snails and leeches are all fairly easy to tell apart in most cases, but the wriggling larvae can be a lot more difficult. The pictures should help, but here is a brief summary of what to look for. It comes as a series of questions, which you should try and answer in the order given.

Q.1: How big is it? In particular, is it larger or smaller than 30 mm? If it's smaller, go on to Q.3 straight away, if bigger go to Q.2.

Q.2: Does it have an impressive pair of scythe-shaped fangs at the front end? If so, it should have a segmented body, clearly visible legs, and no sign of wing pads. Almost certainly this is the larva of a great diving beetle (Dytiscus) but if it's sluggish with very small legs it could be the larva of a great silver beetle (Hydrophilus).

If, on the other hand, it has no obvious fangs but big eyes, wing pads and a segmented rear end it is probably the larva of a hawker dragonfly.

Finally, if it is long and maggot-like, probably segmented, no proper legs and sometimes with a long tube at one end, it's the larva of a true fly.

Q.3: Is it in a case of some sort? If it's carrying one made of plant bits, tiny stones, or sand grains, it's a caddis larva; if the creature is in a cell under a floating leaf, it's a china mark moth caterpillar; if it's not in any sort of case, go on to Q.4.

Q.4: Does it have scythe-shaped fangs at the front end? If so, it should have no wing pads, and go to Q.5: if it doesn't have scythe-jaws, try Q.6.

Q.5: Does it have feathery gills down the sides of its body? If not, it will be one of the smaller beetle larvae (or maybe even a young Dytiscus or Hydrophilus *larva*).

If it does have feathery gills, does it have a single pointed extension to its tail end? This will be an alder-fly larva (Sialis). If not (i.e. if it has no extension, or two or more) it will be one of the few beetle larvae that have gills.

Q.6: Does it have feathery or small, plate-like gills down the body sides? If so, it should also have wing pads and three protuberances (that may or may not be feathery) from its tail end. It's a mayfly nymph.

If it has no obvious gills, does it have wing pads? If not, it is probably either another type of beetle larva in which the jaws aren't obvious, or, if no legs are visible, it will be maggot-like and the larva of one of the true flies.

If it has no obvious gills, but does have wing pads, is it stout and squat-looking with only very short posterior protuberances and large eyes? If so, it's the nymph of a darter dragonfly. If on the other hand it's very slender, often green, with three feathery appendages at the tail end (in other words, much like a mayfly nymph without the gills), then it's the larva of a damsel fly. Finally, if it has two slender protuberances from its back end (and was probably found under a stone in a lake, river or stream) it's the larva of a stonefly.

This key should work most of the time, but will only get you to the family of insect in question. Going beyond, to species level, is infinitely more difficult in most cases and will require a specialized key (see p.122).

Spiders, mites and crustaceans

Insects are not the only arthropods to inhabit our ponds, though they do include the greatest variety of shape and size. There is in fact only one truly aquatic spider, *Argyroneta aquatica* (the water spider), that spends its entire life beneath the water surface. And a fascinating spider it is too. *Argyroneta* spins its web in water plants, not to catch its prey (which it simply runs down), but to hold an 'air bell', that it can then use as a safe breathing chamber and in which it will lay its eggs out of reach of most

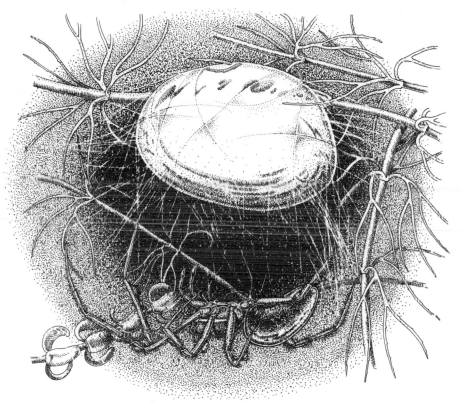

Water spider below air bell (twice natural size).

potential predators. Breeding takes place in spring, and females guard their progeny in the air bell until they are old enough to disperse later in the summer. Of moderate size (big ones spread across 25 mm/1 in or so including legs) and distinctly grey when out of water, beneath the surface it moves shrouded in a silvery bubble of air. Trapping air on its body hairs and swimming down to its web is the way it fills the bell. Another unusual feature of *Argyroneta* is that males grow bigger than females, the opposite situation from most spiders, and presumably making mating rather less traumatic than in other species since dad doesn't face the prospect of becoming mum's next meal. Water spiders are widespread and locally common; they have a wide tolerance of water quality and I have found them in the richest ponds and in pools so poor and acid that little else can survive there.

Apart from properly aquatic spiders, there are several species that frequent the water's edge and a few that venture out on the surface like pond skaters. Of these so-called raft spiders, the giant *Dolomedes* is definitely something for arachnophobes to avoid. This monster (there are actually two species, but they are difficult to tell apart) is very locally distributed, and covers a good 50 mm (2 in) from toe to toe; it's a meaty job too, capable of diving beneath the surface to catch prey as large as sticklebacks. After mating, females carry their eggs around in a large (10 mm diameter) silken cocoon slung beneath them until they hatch and the

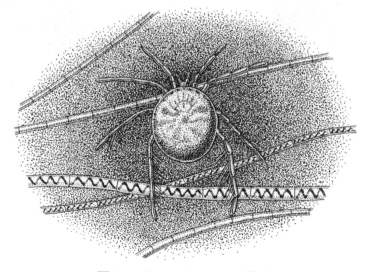

Water mite, ten times magnified.

youngsters leave home. *Dolomedes* is said to have a mildly poisonous bite, and is very much the sort of creature best admired from a distance.

Mites are like tiny versions of spiders, usually just a couple of millimetres across and often brightly coloured. A red variety is particularly common in ponds, and can be found swimming frantically among the weeds. Its larvae, also red like even tinier versions of the adults, suck the blood of aquatic insects and can sometimes be seen attached to water beetles or water scorpions. It's difficult to think of a more obscure group of animals than water mites, but in that great tradition of British natural history it turns out that a complete 3-volume treatise has been written exclusively about them. None of these aquatic members of the spider family can supposedly do people much harm, but I must admit I would think twice before handling a *Dolomedes* and my son once received an unpleasant bite from an exceptionally large *Argyroneta*.

All of which leaves the crustaceans. These champions of the sea (crabs, lobsters and so on) are relatively lacklustre in freshwater, but they do have a few important representatives. The water or hog louse (*Asellus*) looks much like a wood louse, and is common in the smelliest bottom mud of almost all ponds; the freshwater shrimp or side-swimmer (*Gammarus*) is just like a miniature marine shrimp, drab grey-brown in colour and much preferring clean, well oxygenated water. Again, it's a very common little animal whose presence usually indicates rich, unpolluted ponds or rivers. Trout are particularly fond of these shrimps, and a pigment they contain (carotene) is responsible for the salmon-pink coloration of trout flesh from England's finest chalk streams. Smaller still are those well known fish foods *Daphnia* and *Cyclops* and relatives, the so-called water fleas, which, at just 1-2 mm, need close examination to reveal their presence. What they lack in size, however, they often make up for in abundance and in spring ponds often 'bloom' with swarms of these minute herbivores grazing happily on the year's new crop of algae. They in turn are feasted upon by fish, newts and their larvae, and much else besides.

Apart from these common crustaceans, there are also some specialized and much rarer species that are adapted to live in ponds that dry out in summer (see Temporary ponds: a special case). And then there is the crayfish, by far the most impressive member of the freshwater crustacea but rarely if ever found in small pools or ponds. Our native species (*Austropotamobius pallipes*) has been supplemented and partly replaced by various alien introductions, and into the bargain ravaged by a virulent disease which has wiped out populations in many parts of Britain. Crayfish are of course like miniature lobsters, and grow up to 15 cm (½

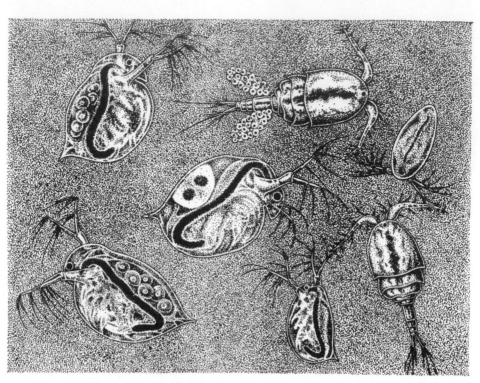

Daphnia *(no tails) and* Cyclops, *both ten times larger than life.*

in) or so in length. Like many other crustaceans, females carry the developing eggs in a mass beneath their bodies and in this condition are said to be 'berried'. The eggs hatch while still on the female, and the young may be carried around clinging on beneath her for some time afterwards. They do crop up in lakes and sometimes in weedy canals; one way of finding them, if you are sufficiently bold, is to stick your fingers into holes or cracks in underwater banks, and wait for the distinctive pinch. The tail is designed for rapid escape, backwards, from would-be predators and if you turn over a rock with crayfish underneath the only sight you are likely to get is of a high-speed reverse flip.

Small is beautiful: the real power of magnification

Many of the pond's most attractive inhabitants are just that bit too small for the naked eye to appreciate, while others are so tiny that they can't be seen at all. It's difficult to imagine how Anton van Leeuwenhoek, a Dutch dry goods merchant and part-time janitor of the seventeenth century, felt when he turned his newly discovered microscope lens on a drop of tapwater and called to his daughter, 'See what I see, Maria!' Within minutes this pioneering spirit must have found more species new to science than all the explorers of Africa put together. Today's investigator of the microbe world has an impressive range of precision instruments available that would be the envy of van Leeuwenhoek and his chums, but they don't come cheap and for this reason alone microscopy will always be at the fringes of pondhunting activity. But if this really is your forte, there are specialized suppliers of good equipment that regularly advertise their wares in wildlife magazines. Basically there are two types of machine to consider: low-power (from ×10 to ×40) binocular microscopes, which are excellent for picking out the minute details on a beetle's leg that are often critical for identification; and medium-power (×100 to ×400) ordinary light microscopes, preferably with phase-contrast optics, that will show the glorious details of diatoms, desmids, protozoans, rotifers, water fleas and so on and even reveal the hordes of tiny, ever-present bacteria. All this is of course work for the table-top back home; microscopes are not field instruments, and decent ones come with their own light source and thus need to be near a mains electricity supply.

Other animals: softies and hard-cases

Water is the ideal medium for having no legs at all, and still being able to get about by exploiting its lubrication and buoyancy properties. Animals that do this include various worms, leeches, snails and mussels as well as more primitive creatures like sponges and hydroids. Many of these are conspicuous, even dominant members of pond communities and are likely to appear in the first haul of the pond net.

That foul-smelling bottom mud is home to most of the true worms that live in freshwaters. Pale, thread-like nematodes often abound if you have the patience (and inclination) to sift through it; but they are all less than 5 mm long, and not usually obvious. Better known are the red worms of the *Tubifex* family, often sold in aquarists shops as live food for pet fishes. Tubifex are especially common in polluted waters, where their pink or red bodies stick up out of the mud and sway about like a field of grass in the wind. They are there to clean the place up, and like most true worms obtain their food from the decaying organic matter all around them. More sinister are the flatworms (*Platyhelminths*), which come in a variety of colours (but most often black or white) and glide effortlessly around in search of living food. Sometimes ponds have plagues of these little hunters (most are less than 15 mm/½ in long) and I have seen frogspawn emptied of its eggs by flatworms crawling all over it and diving in through the jelly. Most fearsome of all, however, are the leeches, of which we have getting

Flatworms among the weed, twice life size.

on for twenty species in Britain; generally thought of as bloodsuckers, many in fact just devour their prey directly and some of the smaller ones are so common it's unusual to find a pond without them. Leeches contract into shapeless-looking blobs when alarmed, but on recovery will rapidly extend to full length; all have suckers on at least one end, and some on both. Leeches get about by looping from one sucker to the other if they have two, or by swimming freely; some regularly come out on land when the surrounding vegetation is wet enough to suit them. Most of the commonest types grow up to 30 mm (1 in) or so, and are yellow-brown or greenish in colour, but two come much larger: the horse leech (*Haemopis sanguisuga*) can be 120mm (4½ in) long when fully extended, and the medicinal leech (*Hirudo medicinalis*), bigger still. Horse leeches are dark brown and fairly widespread; despite their fearsome size they don't attack horses or any other mammal but content themselves with invertebrates and the occasional assault on a tadpole or newt. In fact the medicinal leech, which is now very rare, is the only species (apart from another large and extremely rare one only recently discovered in Britain) where it's necessary to keep a safe distance between it and your bare skin.

Greener than horse leeches, usually with lines of red or orange markings, the medicinal leech was widely used for blood-letting (to relieve black eyes, for instance) until the last century; indeed, some clinics have started using them again recently, and there is now at least one medicinal leech farm in Britain. One thing that makes medicinal leeches different from all other pond life is the way to find them; it's very much a case of 'don't look for us, we'll come to you'. Just splashing about in a pond where these creatures live will, after a few minutes, have them gliding ominously towards you. Keep your eyes open, and most certainly your wellies on; the leeches will stick firmly on your boots trying, of course, to pursue their vampire desires on the tasty goods within. In the unlikely event that one does latch onto you, the best bet is just to sit quiet and let it feed rather than detach it halfway through the meal. The 'bite' doesn't

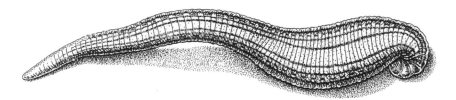

A medicinal leech – real life size!

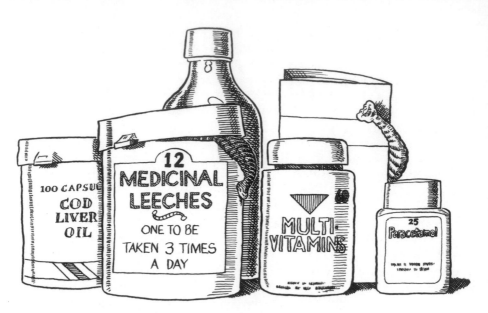

A little bloodletting can be good for you . . .

hurt at all, but when it finally drops off the tiny wound bleeds profusely for a while because of a chemical the leech produces to prevent blood clotting. It was once thought that medicinal leeches relied entirely on birds and mammals to satisfy their sanguine appetites, but it's now known that they are just as content to drain amphibians (which, being much smaller, can be killed as a result). The increased recording of medicinal leeches in the Dungeness area over the past few decades might even be due to their benefitting from the introduction of marsh frogs there in the 1930s; if so, this would be a remarkable example of the introduction of an alien species having a positive effect on a native rarity rather than the usual gripe of foreigners being nothing but trouble. Although rare, medicinal leeches are widely distributed and still occur in places as far apart as Sussex, the New Forest and the Lake District. Leeches are hermaphrodites, each one producing eggs and sperm, so any individual can mate with any other. They do it by intertwining and injecting each other with sperm, which then migrates inside the body to find the ovaries and thus fertilizes the eggs. They put their eggs into home-made cocoons, which may be glued to stones or plant stems in water but are sometimes left out of water altogether in damp mud nearby. Leeches are surprisingly adept at moving over land, and on numerous occasions I have found big ones halfway across the lawn in broad daylight. Birds seem to leave them alone in this apparently vulnerable situation and they don't seem to have many predators in the water either, so it may be that they taste pretty nasty. One family, however, broods both eggs and newly hatched young in a cavity formed beneath its own body. Parental care comes as something of a surprise in such primitive creatures, but is probably a lot commoner than naturalists once supposed.

Often the commonest of all pond creatures are the water snails. Again there are many species, but only two basic designs: the true pondsnails (such as *Lymnaea* species) and the ramshorns (*Planorbis* species). *Lymnaea stagnalis* is a big snail, often with a shell up to 50 mm (2 in) long, and one of the most widespread and abundant of the lot; *L. peregra* is another type often met with, smaller than *stagnalis* and with a much bigger aperture relative to total shell size. As its name suggests, *peregra* is prone to wandering about in damp vegetation and is a quick colonizer of new ponds. Ramshorns are well named, and also come in many sizes; *P. planorbis* is a common small variety, up to 15 mm (½ in) or so in diameter, but the great ramshorn (*P. corneus*), which gets up to a massive

Pondsnail (left) *and giant ramshorn snail* (right), *both twice life size.*

35 mm (1½ in) across, is not at all rare. All water snails feed mainly on dead plant material and debris, but none is averse to the occasional live newt egg and thus can be as predatory as their paltry turn of speed permits. Snails are hermaphrodites like leeches and mate by sticking together for a while. They then lay masses of eggs in colourless, jelly-like capsules which you will find with ease on submerged water plant leaves, sticks, stones and so on throughout the summer months. Adults of some species seem to die off during summer time, to be replaced by the new generations that grow to full size before the autumn. As with most freshwater invertebrates, winter is spent inactive in the bottom mud.

Freshwaters also have so-called bivalve molluscs, with two separate bits to their shells rather than the single casing of 'gastropod' snails. These bivalves are the mussels: the largest, which may even contain pearls, live mainly in deep rivers, but a few are found in the bottom mud of the richer ponds. Swan mussels (*Anodonta cygnea*) are probably the commonest of these, growing up to 150 mm (6 in) long but usually coming smaller except in lakes. Mussels are filter-feeders, sucking water in through a siphon, removing any tasty bits and pieces and expelling the now purer water out the other side. Even so, mussels aren't so harmless all their lives. Eggs are hatched and larvae brooded safely within the shell during the winter months, but these are released in spring as parasites with sticky tails that latch onto any fish unfortunate enough to be passing by. A cyst forms rapidly around the fish's unwelcome guest, which feeds on blood and other body fluids of the host for a few weeks before dropping off to start life as a tiny but proper mussel on the pond bottom.

Much more often found are their minuscule relatives, the orb mussels (*Sphaerium species*) and pea mussels (*Pisidium species*), most of which never make more than a 15 mm (½ in) diameter. They often abound in pond mud, and have the habit of clamping themselves (in the case of orb mussels) to the toes of passing animals. I have seen newts weighed down with orb shells to the point where they can scarcely swim to the surface for air, and toad toes adorned with them like macabre jewellery. All of these tiny mussels filter their food from the surrounding water, and aren't trying to harm their hosts by latching on to a passing free digit; presumably the plan is to secure passage to another pond, though the motives of these miniature mussel hitch-hikers really remains a matter for guesswork. At least their reproduction is less anti-social than that of their larger cousins, missing out as it does that unpleasant parasitic stage.

All molluscs, snails and mussels alike, need plentiful supplies of calcium to build their shells. As a result they are usually most numerous in hard waters; at the other extreme very mineral-poor or acid ponds often have none at all. Despite its obvious protective function, the shell often fails and water snails regularly fall prey to fish, leeches (which worm their way inside) and of course to visiting birds. The larva of at least one water beetle even specializes in them (see Bounders, bugs and beetles), with jaws like miniature can-openers.

This just leaves one other major animal group to consider, the so-called coelenterates. Like crustaceans, these are much better known in the sea than in freshwaters since they include sea anemones and jellyfish; in ponds we have just a few very small representatives, known as the hydras. You won't see these in the pond or the pond net, because, like leeches, they usually contract into tiny blobs when disturbed and in any case are very small and slender (usually no more than 15 mm/½ in long) even when fully extended. However they are widespread and abundant, and the way to see them is to put some pond water and weed into an aquarium and then look at what is sticking to the glass sides after an hour or two back home. Hydras are brown or green, the latter kinds coloured by algae that live deeply embedded in them in a symbiotic ('mutual support') relationship. They all have stalk-like bodies and a head end from which wave varying numbers of tentacles. Hydras are dangerous predators if you happen to be a water flea, or something equally small; those tentacles

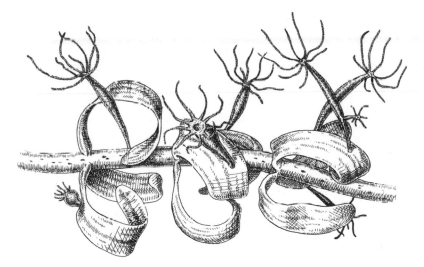

Hydra on pondweed, about twice life size.

harbour deadly poisonous stings, which are shot out like harpoons on the end of tiny threads to stab a passing meal. Hydras reproduce mainly by 'budding'; new animals simply grow out from the bodies of their parents, eventually splitting off to become free-living. Sex rears its head only in the autumn, when the hydras suddenly produce eggs and sperms (sometimes both from the same animal, as some worms and snails are also wont to do) to generate fertilized eggs and subsequently free-swimming larvae that move off to develop into new hydras someplace else. They do this in order to tide them over the winter when there's no food about.

A final word about sponges. These organisms don't look like animals, but animals they are, and not all live in the sea. A few species can be found in ponds and lakes, and you won't need me to tell you what a sponge looks like. They form encrustations on stones, underwater tree roots and so on, but usually only in the larger ponds with clean (unpolluted) water. Like mussels, they obtain their food by filtering out suspended particles of matter from the water around them. Colours range from dirty grey, through yellow to green and shape is so variable I won't even try to describe it.

Microbes: the pond's policy makers

As Jonathan Swift once said, 'A flea hath smaller fleas that on him prey; And these have smaller fleas to bite 'em, And so proceed *ad infinitum.*' It's an undisputable fact that all the life forms big enough to be seen are totally dependent on the proper workings of ones that aren't. Ponds teem with micro-organisms which, without a microscope, no-one would ever guess existed and some are quite exquisitely beautiful. Put a drop of pond water on a microscope slide and a totally new world comes into view, every bit as complex as the one we are normally familiar with.

Tiniest of all are the bacteria, which thrive in the bottom mud, on plant leaves, and even in the open water itself. A typical pond will have more than 100,000 in one millilitre of water, which will still look crystal clear when held up to the light! There are of course innumerable species, but most obtain their nutrition from decomposing animal and plant material (and carry out an essential recycling function while doing so) whereas a few cause diseases in pond animals and others photosynthesize like plants. These 'blue-green algae' can become so dense in polluted ponds that they form smelly mats of very unpleasant-looking gunge. This happens occasionally in reservoirs and is a serious danger sign, because some varieties produce poisons so strong that animals (or people) drinking the water become ill or even die. Mostly, though, microbes remain at relatively low numbers in balance with the many bigger creatures (water insects, snails and so on) that eat them. The presence of the common bacterium of the human intestine, *Escherichia coli*, in freshwater is a sure sign of sewage pollution because this bug can't survive for long outside in the big wide world; looking for it by careful microbiological techniques (it can't be told apart from other bugs under the microscope) is therefore one of many standard ways of checking water quality. Bacteria are just about visible under high-power microscopes, measuring up at less than one hundredth of a millimetre, but a recent surprising discovery is that even more numerous are some tiny viruses which infect and kill bacteria! These are too small to be seen by anything except the powerful electron microscope, but it turns out that your average millilitre of pond water may have as many as 100 million of them, making bacteria seem positively rare by comparison.

Other common micro-organisms, bigger than bacteria, are the true algae and protozoa. Algae come in a huge variety of shapes and sizes, and are of course mostly green since they obtain their food by photosynthesis just like higher plants. Some are immobile, and are just swept around by water currents; others have tiny beaters, 'flagellae', which propel them along, so not all plants have to stay where they are born. Members of one group, the diatoms, have a hard shell made out of silica, the same substance that forms sand grains. Needless to say these are pretty resilient, and many algae can pass through the digestive systems of higher animals like snails or tadpoles and emerge at the other end none the worse for the ordeal. Other algae form colonies that can be beautifully structured, like *Volvox*, or big enough to be seen by the naked eye, as with the filaments of blanket weeds such as *Spirogyra*. Algae, bacteria and detritus form the food of many protozoa, mobile single-celled creatures that patrol the pond bottom and also come in a range of shapes and sizes. Mostly clear or

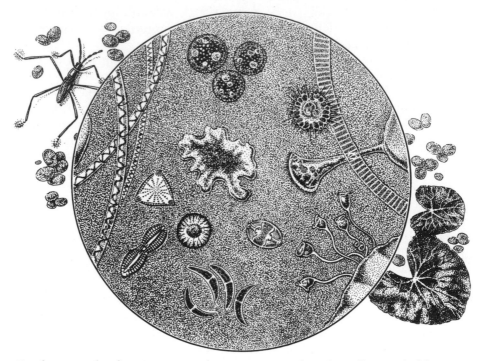

Pondwater under the microscope: algae, protozoa and rotifers, all magnified by 10 or more.

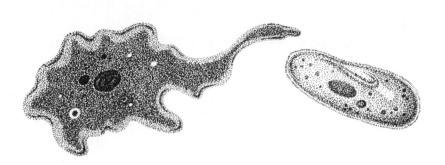

Amoeba (left) *and Paramecium* (right), *both magnified by 200.*

creamy white, well known examples are the amoebae and Paramecium. Protozoa get around by complex changes of body shape ('amoeboid movement') or by using paddle-like structures that may be few and large (flagellae) or many and small (cilia, as employed by Paramecium). Protozoans also demonstrate staggeringly beautiful microscopic structures, but also include sinister members that act as parasites on larger animals including, in tropical countries, man. Fortunately our native Brits are a harmless lot from the purely human point of view.

Although there are some microscopic fungi that live in freshwater, this family is not usually a notable feature of aquatic habitats. One exception is the threadlike *Saprolegnia*, which spreads rapidly like a kind of gruesome cotton wool over animals that perish underwater. This fungus thrives when the weather is unusually cool, and in cold springs it can also be seen growing over frog or toad spawn which often dies as a result.

Finally among the common microbes, and more complex than the rest, the rotifers deserve a mention. Although most are similar in size to protozoans (that is, about 0.2–0.4 mm long) the largest rotifers can reach 2 mm and thus are just about visible to the naked eye. All are made up of lots of cells and have distinct structures including a head, a stalk, and a foot. There are more than 1,000 species known, and probably lots more to find. Being at the 'top end' of the size scale, they are relatively low in abundance; one millilitre of water may contain five of them in a good rotifer pond. Rotifers are predatory, using cilia in this case to propel food items (tiny bits of debris) into what passes for a rudimentary mouth. The microscope, then, reveals a jungle within a jungle; life and death struggles played out in numbers which, in just a few millilitres of water, surpass what goes on in all the great plains of Africa put together.

Waterborne diseases: what can we blame on ponds?

When King Alfred took refuge in the marshlands of deepest Somerset, he was in one sense jumping out of the frying pan into the fire. Standing water has many interesting beasts to tantalize the modern-day pondhunter, who enjoys a relatively safe pursuit; but in days of yore the wetlands of England were a fertile source of the 'ague', or malaria as we call this debilitating disease today. Malaria is a protozoan parasite carried about by mosquitoes of the Anopheles *family; these mozzies still live in Britain, but for reasons that aren't entirely clear (but probably stem from draining the* Fens, and thus reducing drastically their breeding grounds) the malaise they are so famous for has departed our shores. That's not to say it couldn't come back, especially if the greenhouse effect really warms things up, since temperature also seems to be an important factor in the success of this nasty little parasite.

Polluted water is, of course, as dangerous as ever as a potential source of murderous diseases like hepatitis B, dysentery, cholera and typhoid; no-one in their right mind goes pondhunting in sewage, so for most of us these diseases too are problems for historians or public

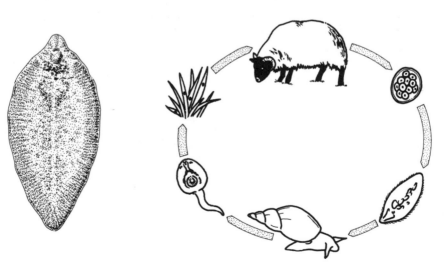

Liverfluke (left) *and a plan of its complex life cycle.*

health watchdogs to worry about. However, freshwaters are not without some slight risks even where everything looks fine and dandy. Weils disease, an infection released into the water by rats, is potentially fatal to humans and claims some victims most years in Britain. The main danger comes from swimming in, and accidentally imbibing, contaminated water and the casual naturalist should be pretty safe.

Probably the most serious health hazard from ponds in recent centuries has been not to people, but to the livestock many ponds were originally made for. Liver fluke is a disease which, in some of the wetter parts of Britain, has achieved considerable economic importance from time to time. This little parasite looks a bit like a flatworm, and has a complicated life history with some stages spent inside water snails and others in large mammals. Sheep or cattle which eat damp grass can take in cysts of this worm, which hatch in the gut and infect the liver causing massive liver rot and thus very serious disease. Even humans, if they are very unlucky, can go down with liver fluke. Don't be tempted to decorate your salad with wild-caught watercress, because it too can harbour fluke cysts and they aren't easily killed or removed by washing.

I suppose from the pondhunter's viewpoint the message must be: take sensible precautions regarding hygiene (wash your hands between netting and eating the lunchtime sandwiches) and don't dabble in pondwater with open wounds or cuts on your hands, but otherwise don't worry overmuch. Almost any other hobby, apart perhaps from stamp collecting, is equally or more dangerous, but take extra precautions in foreign parts which, especially the hotter ones, sometimes host a greater variety of nasties lurking beneath the weed.

Strangers on the shore

Not everything you find in a pond these days is necessarily a true Brit. Aliens have arrived over the years, and in a few cases done well enough to become common (in some cases all too common) members of our freshwater communities. Canadian pondweed (*Elodea canadensis*) has already been mentioned in the plants section (see p. 37), and stands out as the most successful introduced plant species so far. In 150 years it has spread throughout the land, though early worries that it would smother all our native pondweeds were evidently misplaced. In fact it seems to have declined in recent times, after a wild extravaganza last century immediately after its escape. Even so, it remains a common find in ponds and ditches everywhere. The new worry is another water weed still widely sold in aquarists shops, the New Zealand stonecrop (*Crassula helmsii*). This attractive little plant is turning up in ponds all over Britain, especially shallow ones with sandy bottoms. It poses a real threat to rare plants that live in this type of pond, and is very difficult to control; even the tiniest fragment, carried on a wellie boot (or by a bird) can start a new plant in a previously uncolonized pond. There is a national recording scheme monitoring the spread of *Crassula* and a research programme into ways of eradicating it, but so far the weed is winning by miles. Other introduced plants, such as the water fern (*Azolla*) from North America, have spread but not so dramatically as the two main newcomers.

On the animal side there are also a few recent additions. At least three alien species of crayfish, some from across the Atlantic, have escaped from farms and are spreading in lakes and rivers around the country. Apparently there is also a small freshwater jellyfish, about 2 cm (¾ in) across, turning up from time to time; nobody even knows which continent this beast came from, let alone how it got here. Several foreign amphibians have gained toeholds in Britain, but the ones you are most likely to meet are members of the so-called 'green frog complex', thus named because the three types (edible, pool and marsh frogs) are difficult to tell apart. They are all bigger than common frogs, usually at least a bit green, and very aquatic, spending most of their adult lives in or near the water. You may come across them sitting on the banks of ponds or ditches on sunny days in the south or east of England, soaking up the warmth and leaping into the safety of the water with a loud 'plop' when you get too close. The males also croak much more loudly than our native frogs, and generate a fine chorus on spring and summer nights. The first green frogs, very common species in much of Europe, probably turned up in Britain shortly after the French Revolution – not as refugees, but as potential additions to gourmet menus. In some places, particularly low-lying ditches in Kent and Sussex, they have spread widely and become locally very abundant. Fortunately, they don't seem to have had any adverse effects on native species so we can still welcome them without serious concern.

The only foreign mammal you might meet in or around the pond nowadays is the North American mink (*Mustela vison*), the first of which escaped from fur-coat farms in the West Country some fifty years ago. They are now common and widespread, hunting usually in the vicinity of rivers and lakes but making pond visits too in pursuit of fish, frogs or just about any other meaty item they can catch. Mink are quite cheeky and often sit up to take a good look at you before running away; they are much smaller than the now rare otters, and easy to distinguish from them by their size and much darker coloration. Despite intensive trapping efforts there now seems little chance that mink will be removed from our countryside and, like most other introductions that take off, we have to resign ourselves to living with them. The rabbit, after all, was another introduction but we rarely think of it like that because it's been here so long. However, ponds were also the scene of one rare success story in the eradication of an alien animal. The coypu (*Myocastor coypus*) is a large, water-rat like rodent with a similar history to mink; it too was brought to fur farms from the Americas (South this time), escaped in the 1930s and formed large colonies in the Broads district of East Anglia. Unlike mink, it annoyed the authorities sufficiently (by burrowing through artificial river banks and making them leak!) to warrant serious efforts at extermination. Most of these failed dismally, but the massive assault of the 1980s seems to have done the trick. Britain is now free of coypu, we are told. Ironically, this success might have a down side. Recent evidence suggests that coypu grazing of reedbeds in the Broads was a significant factor in keeping the water open for boating and all the other recreational activities that go on there; we will now have to see whether, in ten or twenty years' time, we find the ministry advocating coypu release to stop our Broadlands turning into reed swamp.

Plonkers at the pond: why do they do it?

At the top of a high hill near my home is an isolated, now more or less defunct dewpond. It's a good way from the road for anyone to walk, as dewponds often are; but attractively displayed on the old pond floor are no less than two rusting car bodies and a decomposing bedstead. Now, we also have a municipal dump nearby that anyone can reach with relative ease and which, one might imagine, would be the sensible last resting place for these incredible hulks. After years of this kind of experience there's no doubt in my mind that ponds bring out something deeply mysterious in the human psyche. A need to plonk things in water, so deeply rooted that travel for kilometres over daunting terrain is of no matter, must be at the heart of it. Ask any conservation corps member with experience of pond restoration tasks, and you will quickly realize that the obvious, large items are merely the tip of an enormous iceberg. Buried in the mud of many ponds must be a pretty complete record of twentieth-century civilization, everything from rag dolls to railway toilet seats. In fact the hobby seems to have a long history, limited only by the smaller number of artefacts in olden times;

Another successful midnight splashdown.

one large pond in Hampshire yielded a sizeable crop of Roman coins when searched in the eighteenth century, and I'd bet there are many more such treasures if we did but know it. I suppose pond junkies must be a furtive lot, it's certainly odd how rarely anyone sees them at it considering the impressive scale of their operations up and down the land. I sometimes imagine how the campaigns are planned, with much pouring over maps into the small hours to pick the remotest pond, loading bedstead on the back and hiking singlemindedly through the night to make 'the drop'. There surely is a completely different subculture out there, an underground fraternity of pondhunters with ambitions all their own. Of course we must condemn them, ponds have enough problems without this unnatural succession; but the most crested newts I ever found at one place was in the hood of an old pram contributed by these dedicated if delinquent enthusiasts.

Drop-ins and fly-by-nights

It isn't always easy to decide whether a plant or animal is a 'pond' species, one that sometimes happens to live near a pond, or one that just makes casual visits. Swamp plants are good examples of this problem. Reeds are obviously pond plants; sedges, well probably; but what about rushes that grow just as often in muddy fields? In the animal world, some species seem to have an affinity with water although we wouldn't by any stretch of the imagination call them pond dwellers. Among the reptiles, common lizards (*Lacerta vivipara*) and adders (*Vipera berus*) are quite often found living in grass tussocks (especially purple moor grass) at the edge of or even in shallow heathland ponds. Of course they are found in dry places still more often, but ponds may be good hunting grounds (especially for the lizard, with so many insects about) and this could be the attraction rather than the water itself. As for mammals, in a dry summer ponds are the obvious places to go for a drink and almost any species can be caught at this activity if you wait long enough in the right place. Just like the water-holes of the African plains, ponds become focal points for quenching thirsts, and, though they won't have lions or crocodiles lurking in the background, they do attract predators as well as prey. Hanging around a pond is a particularly good strategy for watching deer in my experience, though you should plan to be there at dawn or dusk for the best chance. Ponds must also be among the best places to see many of our bat species; again it's insects that are the main attraction, and I've had many pleasant but mildly painful evenings around ponds with clouds of bats circling overhead and emitting their high-pitched squeaks. It is usually difficult (at least for me) to tell which species are about, but our commonest (the pipistrelle) is certainly a regular and so are others including real water specialists like Daubenton's.

Like mammals, a huge variety of bird species visit pools for drinking and bathing and any decent pond is likely to be a good birdwatching site. In the garden I have watched crows, magpies, blackbirds and thrushes learning to catch newts in spring and thrushes also take the opportunity provided by large populations of water snails to obtain an easy meal. Arguably more exciting visitors include herons, regular pondhunters that take a variety of insects and frogs as well; kingfishers pop in to some of the larger pools, and on a real red-letter day you might even see an osprey stopping off en route to wilder parts in the hope of a quick fishy nosh. I

A kingfisher surveys the scene.

once saw one hunting over a sand pit pond on the outskirts of London, so it's not as impossible a dream as you might think. Another bird of prey worth looking out for near heathland ponds is the hobby, a summer visitor that specializes in catching dragonflies when they're in season. The list is almost endless, but that after all is part of the fun.

Tourists: the long-haul pond dwellers

It could reasonably be argued that the English Channel, more than any other single thing, has made Britain what it is today. Since that celebrated occasion in 1066, entry into these sceptred isles has been with the consent of the freeholders or not at all. Or so we like to think. In point of fact we experience massive invasions every year, because to many creatures our English Channel is no more than a drop in the ocean. This is all familiar stuff to birdwatchers, alert

to the spring and autumn migrations, but more surprising is the ease with which apparently much feebler beasts pop across when the feeling takes them.

Butterflies like the Red Admiral abound in England every spring, but none was born here; each year they come, try to breed, fail and die. And this overspill effect is a common feature of insect populations, so we shouldn't be too surprised that ponds benefit from it as well. It is established now that some of our dragonflies, such as the Ruddy Darter (Sympetrum sanguineum) regularly arrive from abroad. Many water beetle species are powerful fliers too, and again some are thought to drop in from time to time when France becomes too much for them. Great silver beetles are a case in point; nippy in the air, they are regularly caught in light traps set for moths along the south-east coast of England and it may well be that some of these characters are of foreign origin. Of course it's difficult to be sure unless you intercept one halfway across the Channel, and since many good beetle ponds exist near the Channel coast in England, the light traps might just be picking up a bit of local traffic.

However, back in the last century, more convincing evidence came in the form of several records of a diving beetle species (Cybister) not found before or since in Britain, but which turned up for a while along the coast of Essex. Cybister is a similar size to our great diving beetles (Dytiscus species) but is differently shaped, with a fatter rear-end; it still abounds as near as the Netherlands, and is distinctive enough to be identified easily. With the greenhouse effect in vogue these days it might perhaps try another invasion soon, so keep your eyes peeled for this one.

Temporary ponds: a special case

A question that might well be asked is: when does a pond become a puddle? In other words, how much water does it take to generate a true pond community. In truth, even a puddle that holds water for more than a week or two is likely to be used by midges and mosquitoes. Their larvae grow fast enough to develop fully before the puddle dries up, and they reap the benefits of having nothing else around to eat them or compete with them. But halfway between the puddle and the permanent pond we can identify another distinct habitat, the true temporary pond. These fill up with rain in autumn and winter, usually to no more than a metre or so deep and often much less; in spring the level falls rapidly, and in anything but the rainiest of summers they will be dry by June or July. Ponds like this crop up in all sorts of places, especially those with quickly-draining soils like sand dunes and sandy heaths. It turns out that some animals have adapted to this kind of pond to such an extent that they survive poorly, or not at all, in more permanent waters. Of all the animal groups that have tinkered with this dodgy-sounding lifestyle, it's the crustaceans that really come to the fore.

Daphnia and other tiny jobs can survive long periods of drought by producing specialized eggs that hatch when water returns, and many temporary ponds teem with water fleas quickly after their autumn refill. More impressive, though, are the real specialists. Fairy shrimps (*Chirocephalus diaphanus*) grow up to 30 mm (1in) or so long, and glide like shoals of miniature submarines over the shallow pond bottoms; these animals grow rapidly through the winter months, and in spring produce eggs that simply won't hatch until they have experienced a spell of being dried up on the pond floor in summer. They are therefore found only in temporary ponds, and are much rarer than they used to be. The eggs also

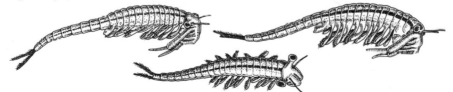

Fairy shrimps, twice life size.

blow around in the wind, and this is how populations start up in new sites; I have seen them in downland dewponds that have cracked and no longer hold water around the year, but an apparently obvious habitat – sand dune 'slacks' – doesn't seem to harbour this species. Even rarer is the tadpole shrimp (*Triops cancriformis*) that can grow up to 40mm (1½ in) long and does indeed resemble a tadpole (or a king crab). It has a similar lifestyle to the fairy shrimp, but in recent decades has been known only from the New Forest in Hampshire, although it's quite common in parts of mainland Europe. All of these beasts have to grow very quickly, and thus require ponds that are not only temporary but also rich in nutrients. Often this is provided in the form of dung by grazing animals coming down to drink, so sifting samples for these rarities is not always the most pleasant of experiences.

It isn't just invertebrates that benefit from the relative solitude of the temporary pond; some amphibians have also learnt that their tadpoles can get a good start in life if they are prepared to risk occasional devastating losses when ponds disappear early in dry springs. In Britain, the rare natterjack toad (*Bufo calamita*) exploits these possibilities to the full. Its favourite breeding ponds are the exposed, sandy-bottomed slacks that form near the sea in the sand dune systems it calls home. In a good year, these ponds are black with tadpoles and thousands of toadlets emerge; in a bad one, the dried-up pond floor is instead black with a sticky goo of dead tadpoles, and nothing at all survives. Fortunately the adult toads are long-lived, and occasional mass infant mortality isn't too serious for them.

Ponds and the war effort

Apart from a little fly fishing by commissioned officers, ponds aren't something of great interest to the military mind in normal times. However, during the Second World War some bright spark decided that the German Air Force might have a few keen pondhunters among its ranks. In particular, the idea was that aircraft flying at night could take bearings from particularly notable ponds by looking for moonlight reflections off the water surface. The answer was, in many cases, to drain what were considered to be the largest and most enemy-useful ponds for the duration of the war. Frensham Ponds both came into that category, being close to large military encampments and depots that were obvious potential targets for bombing raids. It wasn't always easy though; another large pool not far from Frensham was so deep that it proved undrainable, but military acumen provided an imaginative alternative. The entire surface of the pond was covered by felled trees, to such effect that even in aerial photographs taken during the daytime it isn't easy to see where the pond is. No permanent damage seems to have resulted from these activities, and the ponds were quickly refilled and returned to normal in post-war years.

Of course large ponds also provide fun opportunities for training, if you're that way inclined. The same pond that got covered by logs was also the site of experiments with various types of bailey bridge, the remains of which still protrude grotesquely from the water surface. And in the Royal Military Academy at Sandhurst there is (or used to be) an exciting rope slide, from a high tree hide, across a wide and particularly muddy pond that must have been the Waterloo of a few unfortunate cadets over the years.

These days the Ministry of Defence is heavily into conservation on its extensive landholdings, and ponds have been among the beneficiaries. A large and historically famous pond in southern England has been saved largely by the carefully designed 'training exercises' of military

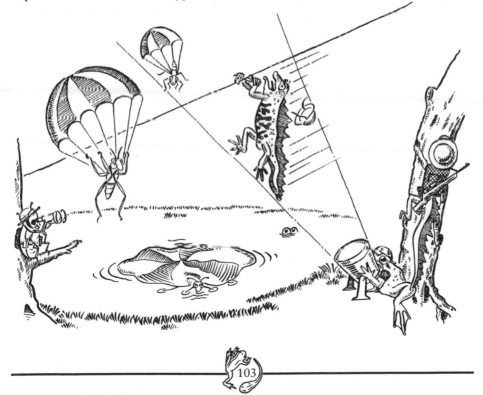

engineers, all part of the positive attitudes to conservation that seem at last to be penetrating the deepest recesses of the Establishment in twentieth-century Britain. Good news indeed.

Ponds at the start of it all?

We'll never know how life began on Earth, nor most of the details of how it subsequently evolved, but ponds feature highly in many of the current theories. Right back at the very beginning, when there was complicated chemistry but nothing yet alive, ponds may have been crucial to getting the right chemicals all concentrated together at the right temperature. We're talking about more than three billion years ago, when one of the biggest problems is thought to have been small chemicals (like amino acids), that could have been formed by lightning passing through a primitive atmosphere, reacting together to make big molecules such as proteins that are the mainstay of all life today. Of course the seas and oceans would have contained these small chemicals, but too cool and dilute (probably!) to react with one another. So think of a shallow pond, perhaps on the slopes of a volcano, where the rain brings in the chemicals after every thunderstorm and then the heat evaporates the water, making ever more concentrated solutions. Of course this is 99 per cent guesswork, but the scene (plus a few other bits and pieces, like a

bottom made out of the right type of clay) is in line with what we know about this crucial chemistry of life. One day all this might be repeated in a laboratory somewhere, but for now it's good science fiction at any rate.

Much later, perhaps a mere 300 million or so years ago, ponds could have been useful again to our would-be ancestors. It's generally agreed that life was stuck in water for all its early history, certainly for 90 per cent of the time since it first appeared. Then, quite suddenly in geological terms, living plants and animals conquered dry land. What triggered this remarkable advance? Again, the scenario of the drying-up pond is often invoked. Picture fishes, perhaps like the African lungfishes of today, gradually adapting to grottier and grottier ponds with mud and not much oxygen. Lungs develop to cope with the dodgy ponds, but one day the water disappears altogether and the fish finds it can live, at least for a time, without it. The first amphibian is born, land is there for the taking and somewhere far down the road this adventurous animal's direct descendants will ponder how on Earth it happened.

Pond ecology: an overview

What we see in ponds is an astonishingly complex web of life, with everything from dissolved chemicals and microbes right through to flowering plants and higher organisms including mammals and birds. Ponds are instructive examples of how the whole living world functions: there is an input of energy from the sun, readily available water (of course!) with dissolved oxygen, carbon dioxide and other important gases, and plenty of dissolved chemicals — all the raw materials that living things need. In such a rich environment it is scarcely surprising that so much activity goes on. Plants, especially the algae, are the primary producers; by using sunlight, water, carbon dioxide and a few simple chemicals dissolved all around them, they are able to grow rampantly and provide food for the next level of organization, the grazing and browsing animals (from amoeba through snails and tadpoles right up to water voles) that eat them. These in turn become prey to the carnivores, in a correspondingly enormous size range (from flatworms to fish as large as pike) and of course there are the various hangers-on in the form of parasites like water-mite larvae, leeches and fish lice. Recycling is catered for by the bacteria and fungi, which convert the results of all this death and destruction back into nutrients and thus bring the cycle full circle. Ponds are lovely places to sit by on a summer's day, but there is a paradox. What is peace and tranquillity to the human observer is, with no holds barred, nature red in tooth and claw to those that actually dwell within.

What make ponds less than perfect places to live, however, are the problems of temperature and time. Being small by definition, ponds are prone to wild fluctuations of temperature that must make life uncomfortable on a fairly regular basis. In summer, water in an unshaded pool can reach blood heat (37°C) or more near the surface, and, unlike lakes or the sea, there aren't any great depths to be plumbed in an effort to cool down. By night a shallow pond can lose this warmth very rapidly, and temperatures may drop 20° or more within the space of a few hours. Many animals can reduce the impact of all this by moving to the most comfortable places, and tadpoles for example often shoal in sunny pond edges during the day but retire to deeper water, that keeps its heat longer, at night; but if you're a plant, or a non-mobile protozoan, you just have to live with it. And in winter there will be ice to put up with, blocking access to oxygen at the water surface and trapping poisonous gases, like methane and hydrogen sulphide formed by decomposition, in the water beneath. All very dangerous, and in severe winters animals hibernating under the ice are sometimes killed in large numbers. Fairy shrimps can be annihilated if their shallow ponds freeze solid, and frogs overwintering in the bottom mud may be suffocated and float to the surface in spring as gruesome bloated corpses.

Time is, perhaps, the pond's most sinister enemy. Ponds are no less mortal than you or I; they are born, suffer the planet's vicissitudes and ultimately perish without trace. Most British ponds in recent centuries have been created by man, usually as watering places for his livestock; natural pond formation is much rarer, but occurs on sand dunes when new dunes form and trap freshwater in the low hollows ('slacks') immediately behind them. Elsewhere it must have happened regularly when rivers flooded their banks, or changed course to leave the so-called 'oxbow lakes', little of which is permitted these days by our tidiness-obsessed river authorities. Even longer ago it's pretty clear that beavers (absent from Britain for at least 500 years now) were first-rate pond-makers; this was an inevitable result of their engineering feats of dam construction, and must have created lots of good freshwater pools in centuries past.

No sooner is a pond made than the processes that will eventually destroy it get underway. Plant growth is the killer; how long a pond survives is probably related to how rich its water is and how many grazing animals are around to keep the vegetation under control, but, with or without this constraint, the plants will eventually win. Deep water gradually succumbs to silt accumulated from generations of dead plants and animals, leaves from the autumn fall and mud and debris washed in by rain; once shallow enough, the reeds and sedges move in fast, converting pond to marsh, and then marsh to dry land. The time scale of all this varies enormously according to local conditions; dewponds on the South Downs were given a maximum lifespan of around a hundred years, but some famous ponds have been around for many centuries (maybe with a little maintenance from time to time); Frensham Ponds in Surrey were created as fishponds for mediaeval monks, and a hoard of Roman coins was found in the bed of another large pond not far distant, suggesting that it was the site of a pond then. By contrast, a dune slack created one year can be filled with sand the next if a heavy winter storm is in the wrong place at the wrong time.

Evidently creatures that live in ponds must cope with the fact that home won't last forever, and may disappear tomorrow. Dispersal is the problem; how to make sure your descendants will find themselves in a new pond should the need arise. With plants, setting seed and casting your fate to the wind is one option but in the pond situation it has obvious limitations: the nearest suitable water may be some way away, and, although many water plants do use this normal plant procedure, a surprising number set seed rarely if at all in this country. Finding duckweed in flower is an ambition most botanists die without ever achieving, while as far as we know all water soldier and almost all Canadian pondweed in Britain are unisex and thus rely entirely on vegetative reproduction — odd bits breaking off and starting to grow as new plants — for their spread and survival.

This obviously works well enough, and may be assisted by the classic 'getting stuck to birds' feet' hypothesis; after all, birds visiting one pond are more than likely to visit another and thus 'target' the bit of weed far better than wind will do for seeds. Amphibians, reptiles, birds and mammals are obviously able to search out new ponds, though some, especially the amphibians, aren't very mobile and quite often suffer local extinction if gaps between ponds get too large. Most insects are able to fly at some stage in their life cycles, and some colonize new ponds with startling speed; I have seen water beetles and greater water boatmen swimming around happily in a new pool within days of its construction. As for the microbes, many pond dwellers are able to form tough cell walls and survive as cysts (cells with hard outer coats) if their pond dries up; these cysts are blown like thistledown on the wind in the hope of finding waters new. This leaves fish, crustaceans and other groups (water spiders and so on) with no obvious means of colonizing new ponds, yet colonize they do; exactly how remains something of a mystery in many cases. Maybe their eggs get blown in the wind, or stick to that weed carried about by birds. Certainly some do better than others, and ponds can exist for many years — often their whole lives — without fish ever turning up in them. So in the long term it is chance, the rolling of the dice, that plays as big a role in the survival of 'permanent' pond inhabitants as it does in the lives of temporary pond specialists on a much more regular basis.

History and folklore: fact and fancy of the village pond

Ponds of one sort or another have been part of England's cultural heritage for many centuries, and it's not surprising that their existence has contributed to story and tradition. Of course most were made to provide sources of water, for drinking and washing, for both humans and their livestock. As such, ponds have been a central feature of rural life since the dawn of civilization in Europe and their importance to the communities around them would be difficult to overstate. Naturally, they developed a significance wider than the original and purely practical ones of quenching thirsts and keeping people clean. I suppose the best known 'cultural' feature of the village pond was its use in punishment. Splashdown on the ducking stool, a kind of see-saw with the unlucky end poised over the pond, was a regular fate for minor offenders in the days before parking fines and probation. Not too serious in summer perhaps, but an altogether different prospect on a frosty January morning. More ominous was the use of ponds during the height of the witch-hunting era, especially during the seventeenth century. Ponds became the original catch-22 for anyone accused of witchcraft; chucked in

bound hand and foot, sinking was proof of innocence (too bad that you had to drown as well), floating meant guilt and execution at the stake. A brief but bloody episode in the role of ponds in human affairs, there are reports of this barbaric practice persisting in country districts until little more than a hundred years ago.

People have long been fascinated by minor mysteries of ponds. Methane, bubbling to the surface during warm weather, sometimes catches fire and tiny, wastrel flames flickering across the pond surface on a summer evening attracted local names such as 'Will O' the Wisp' or 'Jack O'Lantern'. How dewponds get their water has aroused not just interest but passionate argument, with serious texts being published on the subject early in the twentieth century. A feature of dewponds, usually perched high on exposed chalk or limestone hills, was their permanence even in the hottest and driest of summers. They had little 'catchment', and in any case were usually surrounded by porous rocks that water soaked straight into; so how did they keep their levels up so well? Dew was the obvious answer, and it is true that these hills are often shrouded in mist even on summer mornings, but many people couldn't swallow the idea that a bit of hazy water vapour could account for it all. And the sceptics were right; dewponds fill with rain and keep their water simply because they are lined so effectively with clay or concrete, and no downward seepage drains the pond away.

Ponds became, at least for a time, economic resources beyond the obvious one of water supply. It was realized that a suitable water hole was also an insurance against poor harvests and the threat of starvation; after all, ducks like ponds and so do fish. Many large pools, the so-called stew ponds, were created by monks and others in the Middle Ages to keep a handy source of carp for the monastery table. Frensham Ponds, in Surrey, are fine examples. And then there is energy; dam a stream, make a pond behind it and you have the makings of a water-powered mill. Mill ponds of this type exist as relics all over Britain, and some have been maintained or renovated to show another important contribution to village life in the Middle Ages. And of course there were lots of trivial but handy uses too; one, exemplified in Constable's 'Haywain', shows a horse and cart sitting apparently bogged down but really repairing some of the world's first punctures. Wooden cart wheels shrank disastrously in hot weather, and unless reflated by a good soaking there was every chance of an iron tyre pinging off at an inconvenient moment.

More recently, ponds have been created by default: water has filled an old sandpit, mineworking or whatever. Over the past fifty years the image of the pond has changed beyond recognition, from a useful asset to working life to an ornamental essentially luxury item in village and countryside. Fortunately tradition dies hard, and even without economic importance, ponds are valued by plenty of folk. In the 1970s history took a new twist with the highly successful 'Save the Village Pond' campaign; thousands were renovated then and many more since, and with garden ponds coming on stream as well it looks as if a new chapter in our wetland saga has well and truly dawned.

Conservation of ponds: part of our national heritage

Wetlands have had a hard time in Britain lately. There are many estimates around for losses of ponds, most of which indicate reductions of 70 per cent or more within the past fifty years. Serious problems really started with the modernization of farming that took hold after the last war; the money was in arable farming, which doesn't need ponds at all, and even for livestock the trend has been to replace ponds with cleaner, piped supplies to field troughs. The rot started much earlier than this, however; I suppose it could reasonably be dated back to the first successful efforts to drain the Fens in the seventeenth century. There's no doubt that our countryside is a lot drier than it used to be, though a reasonable riposte in the case of ponds is that most of them were artificial anyway; when every field had its corner pond a hundred years or so ago, there was probably more pond life about than this country had seen before or has seen since. The important point, however, is that wetlands and their wildlife are places that many (and increasing) numbers of people value. Hedgerows and Norman churches are human creations too, but we don't accept their destruction on the basis that once there weren't any. We would be wise to ponder, before it's too late, why our ponds are going and what we can do to save what's left.

Of all the perils ponds face today, neglect may well be the worst. Many surveys indicate that most ponds have disappeared by essentially natural processes, coming to the ends of their lives by choking up with reeds and grass (see previous section). Simply by not using them, and hence not clearing them out periodically, we have lost a lot. Others have been drained deliberately to make room for more crops, livestock or houses. The pond that awoke my interest as a boy went that way, under one of Manchester's proliferating estates, sited, it turned out, over a disused mineshaft; the only minor consolation was that it took the developers almost another fieldful of soil to complete their atrocity. Drainage remains a subject of heated argument, but at last the balance in the best remaining wetlands seems to be swinging slowly in favour of conservation.

Memorable battles have been won in the Halvergate marshes of Norfolk and in the Somerset Levels; we need more victories there and elsewhere, but at least they give hope.

Straight destruction isn't the only worry though. Pollution has become a massive problem, mainly as a result of fertilizer and silage run-off from intensively farmed fields. So in many places where ponds have somehow survived the vagaries of agricultural improvement, all we're left with are evil-smelling, bright green pits devoid of all interesting life. Even ponds outside agricultural areas aren't safe; an excellent crested newt pool, within a heathland protected as a Site of Special Scientific Interest (SSSI), was recently filled with stinking pig slurry washed down a field and through a road drain. Other ponds on nutrient-poor soils of heaths and upland moors have been devastated by acid rain, killing off all the fish and sometimes the amphibian populations as well as who-knows how many less visible creatures.

So what is to be done? Two approaches would seem necessary if there are to be any pondhunters around in the twenty-first century, notably the better protection of remaining ponds and the creation of some good new ones to replace all the recent losses. There is undoubtedly an increasing sympathy towards green issues at all levels of society, and many conservation organizations have highlighted wetlands as a priority for action. Ever more areas of outstanding wetlands are being obtained as nature reserves, or at least given the more limited protection of SSSI scheduling; sadly this has sometimes been a painful process, with landowners so outraged by conservationists on one occasion that the press was full of photographs of Nature Conservancy Council officials swinging (as effigies!) from makeshift gallows. Saving large wetland areas is one thing, but preserving the vast network of field ponds on farms across the whole country – every bit as important – is quite another. One option would be to channel some of the somewhat discredited farming subsidies away from overproduction of food and into the conservation of attractive countryside features, thus giving every landowner an incentive to look after his own wet bits. There could even be 'Best Pond of the Year' awards for especially enthusiastic performers. One thing that is badly needed is an inventory of Britain's ponds, so efforts can be channelled into protecting the best of what is left. Hertfordshire led the way on this some years ago, not only producing a comprehensive list of the county's ponds but also grading them from 1 to 5 in decreasing order of interest. Now a National Pond Survey is underway, and with any luck the much needed information on our pond heritage will soon be there for the taking.

Creating new ponds is another valuable option. On nature reserves this has been done to a limited extent for many years, but recently some local authorities have also taken to the idea and on the South Downs restoration of dewponds is very much in vogue. These are among the most expensive to do, because not only does a hole have to be dug, but clay or some other impermeable material needs carting uphill to provide a watertight liner over the porous chalk beneath. Other, rather more accidental successes have come from abandoned gravel pits. When these fill naturally with water they become valuable wildlife havens, though mostly more like lakes than small ponds, and many now exist on the outskirts of London and elsewhere. Old clay- or sand-pits can be even better, typically holding smaller ponds, but local authorities often need persuasion not to use them as convenient landfill sites for the disposal of rubbish.

Ponds are essentially local issues, best defended by the people living nearest to them whose heritage they are. We need more campaigns like the 'Save the Village Pond' one of the 1970s, which proved that local conservation groups can have a real impact, whether campaigning against the latest landfill proposal, lobbying the local council to spend a bit on pond restoration, or mucking in to clear out a neglected, overgrown pond to transform it back to something attractive to the community. In the latter case, a balanced pond should be the objective, not a muddy pool overstocked with ducks and preferably not a repository for all the exotic fish discarded by bored owners; go for a scenic pond, with lilies, irises and plenty of other weeds (and maybe just a few ducks) if you want an attractive feature to help wildlife and put your village on the next 'best kept' calendar.

Newt law: one way to help a pond

Ponds get no special protection in English law, which is one reason why so many have disappeared without trace. However, in 1981 Parliament passed the Wildlife & Countryside Act which, among other things, gave strict protection to the great crested newt and its 'sheltering places'. There were two curious but useful things about all this; one is that crested newts are still widespread in Britain, much commoner than most other species protected by the Act. The newts benefitted from relative rarity in some other parts of Europe, making British populations especially important. The second was the interpretation of ponds, where the newts go to breed, as sheltering places. This term was probably meant originally to mean places like otter holts. Anyhow, the upshot of all this is that there are many thousands of ponds in Britain used by crested newts and which, theoretically at least, are now protected by newt law.

This does have some effect, though, as usual in conservation, it's a compromise rather than a great victory. A massive, multi-million pound housing development in Huntingdonshire would have destroyed a whole clutch of crested newt ponds, but now must be modified to leave some in a piece of land set aside as a newt nature reserve. In Sussex, a road scheme required the infilling of a crested newt pond but this was coupled with the rescue of the newts first, the creation of a new pond some distance away and their transfer to it. There have been a good few such stories lately, though more often what happens is that newts are rescued and dumped somewhere else (which may or may not be suitable) before the pond is lost. Even so, it's obviously a good insurance policy for any pond animal or plant to move in with crested newts if they did but know it. The problems with this are, firstly, that the law is still weak (rescue of animals is no substitute for habitat loss) and that it's so biased towards one sort of pond. Many of the most interesting places I know have never seen a crested newt, so tough luck on them. A better, more comprehensive wetland protection scheme must surely come.

Who else wants the water?

Ponds, lakes and rivers are central to many and increasing human leisure activities, among which the study of their natural history features pretty low in the popularity ratings. Angling of course is a particularly long-standing one with very large numbers of participants. Boating and water-skiing are rather more recent developments, at least on a large scale, presumably reflecting the increased affluence of society in general. Mostly these activities go on around rivers or lakes rather than small ponds, but anglers quite often make do with smallish ponds that in their absence would be quite different. Not only do angling clubs normally want to stock a pond with fish, which as major predators affect (and usually reduce) the wildlife interest, but there may be modifications to the bank and in extreme cases to the water chemistry as well. Heathland ponds, for example, are sometimes limed to reduce their acidity and make the fish grow better. Boating has had a serious impact on parts of the Norfolk Broads, which, as a series of small lakes interconnected by narrow channels, has much wildlife interest more typical of small ponds. The same is true of some old canals, which for many years fell into disuse but are now being resurrected for commercial or leisure purposes once again. Apart from the straightforward disturbance to sensitive wildlife (especially birds and mammals) caused by too much boating or skiing, wave-action by almost continuous boat passage stirs up bottom mud, damages banks and destroys submerged and marginal vegetation. There are pollution problems too, from exhausts, oil spills and even boat-based sewage effluent.

As usual, it's a healthy compromise that's needed and little purpose is served by whingeing on too much about what are undoubtedly legitimate uses of freshwater habitats. After all, these alternative interests have had their plus sides, including the creation of new ponds and lakes and the restoration of old canals that would otherwise have been lost to man and beast alike. This is an area where tact and diplomacy rather than heavy-handed laws should win the day, and with a modicum of commonsense there will be enough water to go round for all of us to enjoy in our various, sometimes devious ways.

Research: taking ponds seriously

Scientific study of ponds, lakes and rivers is much more than the eccentric pursuit of throw-back, nineteenth-century-style naturalists. The quality of freshwater supplies is vital to all of us, and the study of ponds and their inhabitants makes a central contribution to knowledge in this area. The Freshwater Biological Association, with major laboratories in Dorset and the Lake District, carries out important work on issues that can affect all our lives. Many pollutants first register as increased concentrations in ponds or lakes, and sophisticated sampling as well as sensitive measuring devices are needed to warn us of such dangers. Nitrates, which can become dangerous to health when levels rise too high, are a case in point. They enter freshwater mainly as run-off from farm fertilizers, and it is essential to keep a close watch on these and many other chemicals (especially pesticides) if we want to avoid sickness and maybe even the occasional catastrophic poisoning. Many freshwater animals and plants are particularly sensitive to pollution of various kinds, so studying them and monitoring their populations can also provide another useful early-warning system for perils ahead. On a more positive side, research can benefit the commercial use of freshwaters by, for instance, finding out how to obtain maximum yields of plants such as watercress or of fish for sport or the table. And of course there are the delights of study for its own sake, finding out more about how ponds function more or less for the hell of it. All research on freshwater organisms, but especially the 'pure' type, usually has a hard time attracting funds. Even the applied research, giving information on pesticide levels and so on, isn't always welcomed by governments (who might have to do something about it) or industry (who might lose profits because of it). But it all costs very little by comparison with almost any other type of research you can think of, and in this environment-conscious age we should surely be happy to support what is both a long-standing tradition in Britain and an important element of consumer protection. Nowhere more than in this area is research a reflection of a mature and civilized society.

Ponds in the living room: fun with freshwater aquaria

An inevitable temptation, especially when a new or unusual beast turns up in the pond net, is to take it home and keep it for a while. This can be an instructive exercise, but a little thought about when it's a decent thing to do should come first. If a rare animal is involved, the law might be broken (as, for example, in the case of crested newts) by keeping even an accidental catch. Moreover, even if unprotected it may not be fair to deprive a rare creature of its liberty and thus chance making it rarer still. By and large it's better to stick with common species if you want to take samples home for further study.

How not *to bring home your catch.*

How to get the beasties home is another matter of concern. Some pond animals are more robust than others; sticklebacks, and other gill-breathers, won't survive overcrowding or warm water for any length of time and even excessive sloshing about can stun or kill them. Many animals that breathe air (such as amphibians, or water beetles and their larvae) are better carried in containers with damp weed and no water at all. In general, keep your catch as cool as possible and get them home as quickly as possible. Finally, don't mix predators with prey! A pot of tadpoles and newts will quickly become a pot of fat newts only, so if you are planning a 'catch and keep' trip it's always advisable to take several small containers rather than one or two larger ones.

The other matter, which requires altogether more planning ahead, is making sure there is somewhere suitable to keep the creatures when you get home. I've kept water beetles in barrels, boxes and even an old enamel bath but of course by far the best is a proper, glass-sided aquarium. This, filled with tapwater, should be set up in a fairly sunny spot (say, a bay window) with a layer of pond mud and a few oxygenating and/or floating plants. Ideally it should be left like this for several weeks, to settle down and form a balanced mini-ecosystem, before adding anything bigger to it. Naturally the use of pond mud and water plants will introduce small creatures anyway, and it's the growth of these that will eventually stabilize conditions in your tank. Then it will be ready for a few water beetles or their larvae, tadpoles, common newts, dragonfly nymphs or whatever takes your fancy. The same rule about not mixing predators and prey applies to tanks, of course, but fierce individuals such as diving beetles or dragonfly nymphs will need feeding with something. Bits of meat (mince) are often adequate, but take care not to foul the water by adding too much. Many larvae are stimulated to attack by movement, so the meat may need wiggling about a bit. If tadpoles are kept until they turn into frogs, or if water beetle larvae are kept until they are ready to pupate, it is crucial to provide a bank of soft mud that they can climb out onto. Most pond animals will settle down happily in an aquarium, though newts sometimes become restless and males quickly lose their crests if conditions aren't to their liking. The best plan is to keep any particular creatures just for a few weeks, long enough for a good look, and then release them back where they were caught. This is important, because chucking them in the nearest pond might seem OK to you or me, but in fact that pond might be a totally unsuitable habitat in the opinion of the animal in question. A lot of fun can be had by studying pond creatures in this way, and done properly it's both safe for the individuals concerned and good conservation practice.

Garden ponds: the great wet hope

Not all news on the pond front has been bleak over the past few years. One encouraging trend has been the vogue for garden 'water features', mostly it must be admitted of a kind that doesn't help pond life much. Garden ponds are usually stocked with ornamental fish at densities far higher than you would ever find in natural ponds of similar size, and that means lots of permanently hungry predators eating everything else in sight. A wildlife pond aims for a more balanced approach, and to my mind has much more to offer than the formal garden pond set-up. So here are a few tips for the creation of a pond that will be an attractive garden feature, harbour a variety of pond life and even make a contribution towards conservation all at one go.

Where to put your pond, and how big to make it, are inevitably the first considerations. The answers are simple: in as sunny a spot as possible, avoiding trees or bushes that will shed leaves into the water in autumn, and as big as space and money permit. More significant than size is shape: 75 cm (30 in) is a sufficient maximum depth for any pond, and this should be as a deep hollow surrounded by shelves of various lesser depths including a very shallow one of no more than 10 cm (4 in). A good plan is to include a marshy extension to the pond, permanently filled with damp mud. The 10 cm shelf will be used by frogs for spawning, and the marsh will not only harbour an interesting selection of plants but also provide an essential pupation site for water beetles. One problem with artificial ponds is that there is normally a sharp transition, from totally wet to very dry; this is not at all like natural ponds, which usually have damp surrounding soil that many species need sometime in their life-cycles. As for materials, there are a variety of options these days including pre-moulded plastic and fibreglass shells, flexible liners that fit any shape you dig, and good old-fashioned concrete. In the past I've used liners a lot; the toughest (and most expensive) is butyl, but despite the resistance of this material to damage by sunlight it still has its problems, in my experience. Some birds can peck through it, and occasionally do so when hunting newts or water snails around the pond edge; a cantankerous local fox seemed to find one of my ponds the ideal place to exercise its bad temper, by digging up sections of pond edge and ripping great chunks of liner with its teeth. There are even some water plants whose seeds, if they get underneath, grow up with razor-sharp tips and come right on through the stuff (be warned, avoid the burr-reeds that specialize in this destructive growth). Although holes in butyl are fixable, they are often hard to find and then there's all the hassle of draining the pond, drying the surfaces and so on before a seal can be made. So I have reverted recently to concrete; this can be delivered ready-mixed, and though moving the stuff to the pond hole and shovelling it in place is still harder work than making ponds any other way, I am much more confident about its long-term prospects if laid properly (i.e. with outwardly angled, not vertical sides to avoid ice damage and with a thickness of at least 10 cm/4 in everywhere). You will also have to treat the concrete with Silglaze, a cheap material painted on afterwards to seal the pond and neutralize the alkalinity that would otherwise kill animals and plants put in the water. Total cost of a concrete pond is similar to a butyl one, but there are other advantages to concrete apart from being animal- and plant-proof. You can actually wade safely into the pond without fear of damaging the bottom; excess plants can be raked out with abandon, and it is even possible to install plate-glass windows in pond sides and watch the inhabitants just like an outdoor aquarium. I've done this on two ponds, and heartily recommend it. Of course you need to dig a pit down the side of the pond, but this can then be equipped with comfy chair, piped music or whatever takes your fancy. Altogether the ultimate in state-of-the-art pond watching.

Stocking the pond is much like setting up an aquarium. Go for native plants (bought or obtained with permission, of course) rather than Canadian pondweed or the like; they can be contained in baskets full of pond mud and placed on the bottom, and a thin layer of soil should go on all the shelves except the shallowest where coarse gravel looks rather better. Make sure you have a good mix of submerged oxygenators, with floaters in the deep section; these can be native water lilies or the floating-leaved *Potamogeton natans*, but for a true wildlife pond try and resist the temptation of the exotic (and admittedly more colourful) cultivated water lilies. Give over some of the pond edge to marginals, such as irises, reeds or sedges, and in very shallow bays grow water forget-me-not, brooklime or water parsnip. Encourage grass or rockery plants to grow around, and eventually cover up, the exposed edges of concrete or pond liner.

Patience is essential in the making of a wildlife pond; expect it to cloud up with algae, and develop horrendous growths of blanket weed that need regular removal, for at least the first year or so. Eventually it will settle down, as the higher plants become established, and then is the time to go to town introducing animals. Again, make sure you take these from places where there are lots to spare (or preferably from a friend's garden pond) and don't pillage rarities. Frogspawn should go on the shallow (10 cm) shelf, to start your local frog population, and toadspawn can be wrapped around plants in rather deeper (20-30 cm/8-12 in) water. Don't try to introduce adult frogs or toads, they are likely to wander off and probably get killed by something. Newts, on the other hand, can be put in if you know of a place where you can find half a dozen or more smooth or palmates; a big pond can be good for great crested newts too, but this will need a licence from the Nature Conservancy Council.

A few netting sessions in wild ponds should provide you with a variety of suitable inhabitants including snails, water lice and shrimps, water spiders, water beetles, dragonfly nymphs, water scorpions and so on; don't take too many, since overcrowding will cause starvation or emigration in the case of those that can walk or fly away. I would recommend introducing small animals low down the food chain (water lice and shrimps, snails) some months before the predators so they have time to get established and thus won't all get eaten in the first few days. Normally, I think it's best to avoid fish altogether in a small garden pond but sticklebacks do well in larger ones and it is fascinating to watch their territorial disputes, nest-building and so on. The price you will pay is that some other creatures will fare badly in stickleback ponds; newts, in particular, are unlikely to become established as the small fish are astonishingly efficient predators of newt tadpoles. On the plus side, sticklebacks form an all-year-round food supply themselves for some of the most impressive insects like diving beetles and water scorpions, so it really depends on what you most like to see. Of course the real solution is to have two ponds!

Maintenance should amount to little more than removal of excess pondweed each year, an operation best carried out in late autumn when not much else is around. Apart from that, just sit back and enjoy it all. A wildlife pond is for many people an almost magnetic attraction, something to spend happy hours dawdling by. To my mind, few (in)activities are more agreeable.

The pond as a nice little earner

The value of ponds as an economic resource didn't entirely disappear with the advent of modern farming methods. There's still money to be made from the right kind of water body, and still people around that earn or supplement a living from them. In many parts of Europe the significance of ponds and marshes as places to shoot wildfowl attracted down for a perilous drink or bath has not gone unnoticed. I have special memories of discovering that a particular set of cute little ponds in France, which I naively thought of as wildlife havens, were in fact created to be bloody battle grounds for an hour or two each dawn and dusk during the wildfowl season. Hardened veterans blasting away all around makes for a disquieting pondhunt, though I was heartened to notice that these erstwhile shocktroopers always seemed to return empty handed. Angling is another and much commoner pond pastime with commercial bent. Many a farmer has discovered that, for the sake of a few days' work digging a big hole with a JCB and bunging in a few rainbow trout he can expect a queue of patient anglers prepared to pay astonishing sums for the privilege of hoiking the fish out.

But not all modern uses of ponds have followed such obvious paths to fortune. Early this century the interest in water beasties as aquarium pets was catching on, and it occurred to some entrepreneurs of the day that a living might be had on the back of this hobby. The idea was simple indeed; dig a few ponds, colonize them with the more popular animals (frogs, newts and so on) and reap the harvest. One set of such pools in Surrey became quite well known to naturalists, and until recently was a going concern run by a large biological supply company. Apart from the native British pond life, various aliens were either deliberately introduced or escaped from enclosures over the years, and pondhunting in the area is still especially exciting because you really don't know what will turn up in the net. Alpine newts and edible frogs in particular have spread from this patch of ponds to colonize others up to a kilometre or more away. It's doubtful whether much of a living could be made this way nowadays, but a more recent development is certainly commercially viable. Garden centres cater increasingly for enthusiasts setting up their own garden pools, and there are even specialized companies in some parts of England that will design and build a complete wildlife garden for you, with a pond as its central feature.

Societies, books and identification guides

There are now several specialist groups in Britain where you can go for more information on pond dwellers, and of course for the opportunity of meeting like-minded people. The most useful I know of are listed below:

The Royal Society for the Protection of Birds (RSPB) is well known to most folk. The society has many wetland reserves, and its headquarters are at The Lodge, Sandy, Beds SG19 2DL.

The Mammal Society is increasingly active in conservation, and for information on wetland species (or any others) it can be contacted at the Department of Zoology, Bristol University, Woodland Road, Bristol.

The Otter Trust is, as its name suggests, a specialized institution devoted to safeguarding the future of this marvellous animal in Britain. Visit, and see otters in the various captive-breeding and release programmes; the trust is at Earsham, near Bungay in Suffolk.

The British Herpetological Society is active in the study and conservation of amphibians and reptiles in Britain; it produces several leaflets, including one on garden ponds and another on surveying for British amphibians. Contact the Secretary, c/o Zoological Society of London, Regents Park, London NW1 4RY.

The British Dragonfly Society has blossomed since its creation in 1982, and now produces interesting leaflets including guidelines for the identification of British species. Contact it at 68, Outwoods Rd, Loughborough, Leicestershire LE11 3LY.

The Balfour-Browne club caters for keen water-beetlers, of which there are a surprisingly large number for what might be considered a singularly obscure sport. It produces regular newsletters; contact Dr G. Foster, 3 Eglinton Terrace, Ayr KA7 1JJ.

Of course there are also Trusts for Nature Conservation in every county, and joining yours will give access to local knowledge as well as nature reserves in which pond life can be enjoyed. Find it in the telephone directory(!) or via the national headquarters of County Trusts at the Royal Society for Nature Conservation (RSNC), The Green, Witham Park, Waterside South, Lincoln LN2 7JR.

There have been numerous books on pondlife over the years, and, though many are out of print, some are well worth looking out for in second-hand bookshops. Any library worthy of the name should also have a decent selection of them. Particularly recommended are:

The Freshwater Life of the British Isles by John Clegg (Warne, London): there have been three editions with the last in 1965. An enjoyable read from one of the country's most renowned pondhunters.

British Naturalists' Association Guide to Ponds and Streams, again by John Clegg (Crowood Press, Wiltshire, 1985). Very user-friendly, with lots of good photographs.

Life in Lakes and Rivers by T.T. Macan and E.B. Worthington. One of the first of Collins (London) New Naturalist series, published in 1951, it has the authoritative hallmarks of its very distinguished authors. Get a copy if you can!

Collins Field Guide to Freshwater Life by R. Fitter and R. Manuel. A Collins publication of 1986, this is the most up-to-date and comprehensive of all the general books on this subject and has many excellent colour photographs. An absolute must.

Ecology of Fresh Water by A. Leadley Brown. A very enjoyable little book (Heinemann, London, 1972).

The Natural History of Britain and Northern Europe: Rivers, Lakes and Marshes by B. Whitton (Hodder & Stoughton, London, 1979). Very good for a wide range of plant and animal identifications, with useful colour pictures.

There are also a couple of golden oldies:

The Natural History of Aquatic Insects by L.C. Miall (Macmillan, London, 1895)

Life in Ponds and Streams by W. Furneaux by (Longman, Green & Co, London, 1896)

And just for fun:

Water Beetles and Other Things by F. Balfour-Browne. Published probably in 1962 (no date is given!) by Blacklock-Farries of Dumfries. This gives a real insight into life at the shallow end. Difficult to find copies now, unfortunately.

Beyond these general interest books come a number of more specialized ones, aimed at identification of plants and animals towards the species level. A few goodies are:

The Pocket Guide to Wild Flowers by D. McClintock and R.S.R. Fitter (Collins, London, 1955); despite its title and age, for me it is still among the best books around for identifying, and learning about, British water plants. Very comprehensive treatment that covers all species.

The Wild Flowers of the British Isles by I. Garrard and D. Streeter (Macmillan, London, 1983); this has the best colour pictures of water plants I have seen anywhere.

British Water Plants by S.M. Haslam, C.A. Sinker and R.M. Wolsey; (Field Studies Council, 1975); by far the most specialized and comprehensive key to all the different varieties.

A Guide to Freshwater Invertebrate Animals by T.T. Macan. (Longman, Essex, in several editions between 1959-1981). Excellent for identification down to family level for all those beasties crawling about in the net, with useful drawings to accompany the keys.

British Snails by A.E. Ellis (Clarendon Press, Oxford, 1969).

The Dragonflies of Great Britain and Ireland by C.O. Hammond (Curwen, London: 2nd edition, 1983). Excellent pictures and keys.

A Complete Guide to British Dragonflies by A. McGeeney (Jonathan Cape, London, 1986); another first-rate volume.

Caddis larvae by N.E. Hicken (Hutchinson, London, 1967).

Frogs and Toads by Trevor Beebee (Whittet Books, London, 1985). Need I say more?

A Key to the Adults of British Water Beetles by L.E. Friday. An AIDGAP guide (Field Studies Council, 1988)

Collins Guide to the Freshwater Fishes of Britain and Europe by B.J. Muus and P. Dahlstrom (Collins, London, 1978); a complete work with all species the pondhunter is likely to meet.

A Field Guide to the Reptiles and Amphibians of Britain and Europe by E.N. Arnold and J.A. Burton (Collins, London, 1978).

The Mammals of Britain and Europe by G. Corbet and D. Ovenden (Collins, London, 1986); another of the high quality guides from this publisher including of course the various aquatic species.

I have deliberately left out birds, simply because bookshops overflow with good bird books and picking the best is really the business of birdwatchers rather than a true part of the pondhunter's craft. And of course there are some groups, such as various types of fly larvae, for which books don't exist. For these you will need more specialized scientific publications, a handy list of which is given in Fitter & Manuel's Collins guide mentioned above. Finally, there is now a magazine often featuring pond life that is well worth a look; *British Wildlife* is published six times a year from Lower Barn, Rooks Farm, Rotherwick, Basingstoke, Hants RG27 9BG.

Index

If you have enjoyed this book, you might be interested to know about other titles in our **British Natural History** series:

BADGERS
by Michael Clark
with illustrations by the author

RABBITS AND HARES
by Anne McBride
with illustrations by Guy Troughton

BATS
by Phil Richardson
with illustrations by Guy Troughton

ROBINS
by Chris Mead
with illustrations by Kevin Baker

DEER
by Norma Chapman
with illustrations by Diana Brown

SEALS
by Sheila Anderson
with illustrations by Guy Troughton

EAGLES
by John A. Love
with illustrations by the author

SNAKES AND LIZARDS
by Tom Langton
with illustrations by Denys Ovenden

FALCONS
by Andrew Village
with illustrations by Darren Rees

SQUIRRELS
by Jessica Holm
with illustrations by Guy Troughton

FROGS AND TOADS
by Trevor Beebee
with illustrations by Guy Troughton

STOATS AND WEASELS
by Paddy Sleeman
with illustrations by Guy Troughton

GARDEN CREEPY-CRAWLIES
by Michael Chinery
with illustrations by Guy Troughton

URBAN FOXES
by Stephen Harris
with illustrations by Guy Troughton

HEDGEHOGS
by Pat Morris
with illustrations by Guy Troughton

WHALES
by Peter Evans
with illustrations by Evan Dunn

OWLS
by Chris Mead
with illustrations by Guy Troughton

WILDCATS
by Mike Tomkies
with illustrations by Denys Ovenden

Each title is priced at £6.95 at time of going to press. If you wish to order a copy or copies, please send a cheque, adding £1 for post and packing, to Whittet Books Ltd, 18 Anley Road, London W14 OBY. For a free catalogue, send s.a.e. to this address.